The Euthanasia/Assisted-Suicide Debate

Recent Titles in
Historical Guides to Controversial Issues in America

The Gambling Debate
Richard A. McGowan

Censorship
Mark Paxton

The Torture and Prisoner Abuse Debate
Laura L. Finley

Affirmative Action
John W. Johnson and Robert P. Green, Jr.

Alternative Energy
Brian C. Black and Richard Flarend

The Healthcare Debate
Greg M. Shaw

Global Warming
Brian C. Black and Gary J. Weisel

Separation of Church and State
Jonathan A. Wright

Evolution, Creationism, and Intelligent Design
Allene Phy-Olsen

Prostitution and Sex Work
Melissa Hope Ditmore

Capital Punishment
Joseph A. Melusky and Keith Alan Pesto

The Euthanasia/Assisted-Suicide Debate

Demetra M. Pappas

Historical Guides to Controversial Issues in America

AN IMPRINT OF ABC-CLIO, LLC
Santa Barbara, California • Denver, Colorado • Oxford, England

Library of Congress Cataloging-in-Publication Data

Pappas, Demetra M.
The euthanasia / assisted-suicide debate / Demetra M. Pappas.
p. cm. — (Historical guides to controversial issues in America)
Includes bibliographical references and index.
ISBN 978–0–313–34187–8 (cloth : alk. paper) — ISBN 978–0–313–34188–5 (ebook)
1. Euthanasia—Law and legislation—United States. 2. Assisted suicide—Law and legislation—United States. 3. Right to die—Law and legislation—United States. I. Title.
KF3827.E87P37 2012
344.7304′197—dc23 2012022300

ISBN: 978–0–313–34187–8
EISBN: 978–0–313–34188–5

16 15 14 13 12 1 2 3 4 5

This book is also available on the World Wide Web as an eBook.
Visit www.abc-clio.com for details.

Greenwood
An Imprint of ABC-CLIO, LLC

ABC-CLIO, LLC
130 Cremona Drive, P.O. Box 1911
Santa Barbara, California 93116-1911

This book is printed on acid-free paper ∞

Manufactured in the United States of America

Barry T. Bassis variously questioned, harangued, read, and fed me during the course of the PhD and this later project—which was originally supposed to be a "mere" extension of the dissertation's first chapter, with some amplification, and which turned into an exploration of doctor prosecutions, their risk, and their impact upon the historical debate of medical euthanasia and physician-assisted suicide in the United States. On a number of occasions, he suggested that I dedicate this work to him. This I shall not do, as I am too superstitious to so dedicate a book about death; this is especially so, since he very nearly died in the earliest stages of the work, with the result that I am reluctant to tempt fate. Rather, this project was dedicated from the get-go to grandfather Neocles "Pappou" Pappas, who would have encouraged the questions; and to father William "Bill" Vasilis Pappas, who would have encouraged the project and doctoral study had he not been felled by Huntington's disease, or perhaps because he was.

"Journalists write the first draft of history."

—Maryn McKenna, MSJ. Quote excerpted with kind permission from "The Bacteria at the Back of the Garden: Atlanta, the CDC and the Power of Not Being Noticed," paper presented at the Annual Meeting of the Southern Association for the History of Medicine and Science, Atlanta, GA, March 2, 2012.

Contents

Acknowledgments

That researching and writing about the social and legal constructions of euthanasia and physician-assisted suicide found its way into an academic series of projects from doctoral dissertation to articles to book was the result of very encouraging support of my teachers and mentors at the London School of Economics and Political Science. Professor Paul Rock (Sociology) and Professor Robert Reiner (Law) planted a seed that took root over the course of the year in which I earned my MSc in criminal justice policy at the LSE, and continuously nourished the project in the years to follow—with everything from economic support to moral support, from amazingly thorough editorial efforts to excellent questions, from suggestions as to practical advice for fieldwork to readings over a wide and fascinating array. Professor Leonard Leigh (Criminal Law), who ran an LLM course in theoretical and comparative criminal law, encouraged me to make a jump from the practicing lawyer I had been to the academic lawyer I could become. Professor David Downes (Department of Social Administration), in a brief but highly influential conversation we had at the end of my master's year, told me that "anything historical is inherently comparative," which offered me a new, comparative approach to work that had previously been constricted in my professional life to a single case before the court at any one time.

When I became a doctoral candidate, I knew both everything and nothing. I knew how I felt about a topic that I swiftly (and perhaps permanently) became ambivalent about. I did not know how to conduct an academic inquisitorial interview, having been trained to be a lawyer in an adversarial setting. Every person I interviewed, whether quoted or not in the resulting thesis and writings, has my profound gratitude for their time and their patience, as well as for their information. I knew how to read a transcript, but not a roadmap to the courthouses to which I suddenly had to drive (for this, I thank a law school friend, Ann McCloskey, for a nonfinancial night on Wall Street, giving me driving lessons before my first trip to Detroit; I also thank the people of Michigan for their defensive driving skills and explaining to an apologetic woman what road skill she needed at any given moment for months on end).

In a similar vein, my wonderful physical therapist, Avis Toochin, taught me how to walk, encouraged me to run, and inspired me to fly, after Dr. Stuart Springer performed the first of five miraculous operations to repeatedly rebuild both knees between 1990 and 2007. Without these two, I would never have been able to move to London or to drive or to do fieldwork (though I note that Dr. Springer cautioned me, without success, against doing so for the two 1996 Kevorkian trials, which successfully provided much of what will be reported on in the field chapters). Mrs. Hermine Silver, office manager to Dr. Springer's very busy practice, repeatedly ensured that all manner of administrative matters were attended to, so that I could use the time and mental energy to focus my energies upon my academic work, as well as my knee rehabilitation. Dr. Ron Noy performed a final surgery, which literally enabled me to walk at my PhD graduation. In the later period of post-millennium surgeries, I literally had "A-TEAM" in New Jersey, with Dr. Robert Ruffalo and his very able physical therapy staff, including Jan Steele, Walter "the other Rob," and Joanna.

The people (small case deliberate) of Michigan became the focus of the original doctoral study, which went far beyond the original contemplation of the numerous cases of *People v. Jack Kevorkian.* I was fortunate to meet individually with many of those involved in the Kevorkian cases, as well as the Michigan Commission on Death and Dying. These included doctors, nurses, patients, family and friends of Kevorkian decedents, lawyers, legislators, legislative task force members, judges, parties, jurors, witnesses, "named plaintiffs" (parties bringing civil actions against the state), "named prosecutors" (heads of offices, known in Michigan as "Chief Prosecuting Attorneys"), "line" (courtroom and appellate) prosecutors, professors and professionals, secretaries and clerks, ordinary people in extraordinary circumstances, who were all extraordinarily generous with time and information in varying forms.

This project reflects my emerging awareness that the time was right for developing research and writing methods of an interdisciplinary nature. My original training in law came out in particular ways of looking for lacunae, asking questions, seeking information. For this project, this provided some rich opportunities for context and perspective. My subsequent academic training in criminal justice, sociology, and law at the LSE is expressed in this work by an expanded vocabulary, none of which will frighten the American lawyer more than that I learned to ask "Why?" Lawyers became concerned by my questioning of doctors, which prompted me to question nurses, and to focus upon medical personnel in a number of ways, later and most unusually in regard to the *voir dire* of Kevorkian's 1999 trial. I became obsessed with the historical development of the language of medical euthanasia and physician-assisted suicide, and I thank everyone whose writings and comments encouraged (or challenged) me in this regard (and those who will do so in the future).

It is my hope that this project will be of interest to a general readership composed of members of all of these groups, and others who are interested; I pointedly seek to be conversational, while assuming an intelligent, curious reader (to paraphrase the words of Paul Rock). For a number of years, I was adrift in a sea of shifting vocabularies amongst professions and disciplines, and Paul Rock (again) quite rightly commented that I was concerned that I was neither fish nor fowl. Years of my ongoing concern that I would be able to neither swim nor fly led to the conclusion that this may last the rest of my life. Professor Phil Kayal at Seton Hall University paid me the high compliment that I was both a lawyer and a sociologist at a time I was worried that nobody would see me as either. Robert Reiner, a sociologist who later trained in law, kept me afloat in the idea that I could be both interdisciplinary and multidisciplined. That said, any ambiguity or error in any discipline is something I offer an apology for, in the same way that I offer apologies for English (U.S.)/English (UK) shifts in spellings and usages, which changed depending upon what I was quoting or whom.

I am grateful to countless friends, colleagues, collaborators, and advisers for their generous help in many ways and at many times. Similarly, a number of institutions must be thanked.

Formative conversations were golden opportunities over the years. Lunches with friends in New York (sometimes during fly-by visits and layovers) with clerks and judges at the Appellate Division in New York were as spirited as the appellate arguments of my early days of practice, and certainly influenced me in questions I asked during fieldwork or focused upon during writing up the analysis. The Hon. Ernst H. Rosenberger paid me the compliment of listening to my access questions (and offering insightful answers) and the ultimate compliment of discussing his own work in a master's of judicial administration and his late father's doctoral work (regarding a public health matter) prior to the Holocaust. The Hon. Betty Weinberg Ellerin tirelessly cheered me on to do academics, when few supported the decision. The

Hon. Joseph Sullivan read my work and sent brief, pointed, and pointedly encouraging e-mails, which to this day warm my heart. The Hon. John Carro, when I announced that I was leaving to go to graduate school for a year (that became a lifetime of master's, doctoral, and postdoctoral work), gave me the high compliment that if I was going to the London School of Economics and Political Science, I was making a good selection, because his sister had studied there. The late Hon. Theodore Roosevelt Kupferman encouraged me to make presentations to the Legal History Committee of the Association of the Bar of the City of New York and the New York Society of Medical Jurisprudence, while I was working on my doctorate.

After I completed the PhD, Professor John Q. Barrett of St. John's University Law School and Ken Aldous, Esq., in their roles as chairs of the Legal History Committee, invited me to speak about how the unique legal history of Michigan that made the possibility of a multi-tried (under a variety of legal theories and statutes) Jack "Dr. Death" Kevorkian and to branch out into other areas of interest in socio-legal studies. As I began to make the transition from legal practitioner to socio-legal researcher, Professor Glennys Howarth (University of Sydney Faculty of Health Sciences and University of Bath Centre for Death and Society, and currently a dean at the University of Plymouth) and the Rev. Dr. Peter Jupp introduced me to the sociology of death, which became a particularly lively field in the 1990s, and graciously provided a neophyte with first conference, publication, and book review opportunities. Professor Howarth later served with grace and rigor as my external doctoral examiner, and, along with my internal examiner, Professor David Downes, provided me with one of the most productive (and surprisingly relaxed) conversations of my entire life. Unrelated officially to the PhD, Professor David Schultz of Hamline University also helped me to develop by inviting me to write a number of papers along the research route and periodically serving as a sounding board.

Dr. Bill Gasarch (an academic at the University of Maryland) and Ms. Carolyn Gasarch (a trained pharmacist), Dr. Pearl Gasarch (whose PhD was in technical writing) and Mr. Ralph Gasarch (with an MA in drama) routinely offered questions outside the law, and answers about the academy. Dr. Linda Wasserman and Dr. Lewis Wasserman issued medical challenges and challenges about medicine. Hilly Wiese, Esq., provided a number of crucial insights. The Hon. Margaret Marrinan of Minnesota cheerfully kept telling me to "push" at critical stages of the original doctorate. Rob Weinberg offered critiques as to my perspectives and weekly columns to educate and entertain me, along with e-conversations that were almost as good as being there. Professor Mike Lewyn was always a tireless cheerleader. Ms. Lisa Haddock kindly co-taught with me on a number of occasions and sent me many an article of interest.

My flatmates over the years were never-ending sounding boards at all hours of the day and night, and I thank the Women of Flat 6 (and Vasili), the Men of Flat 30 (especially Dr.—now Professor—Elias Mossialos, who encouraged me to apply for a fellowship at the University of Minnesota, and Jonathan Springer, who helped me formulate thoughts into a proposal; I also thank Elias for introducing me to Dr. Dina Davaki, who, with Dr. Francesco D'Amico, invited me to give a seminar on the role of the media in assisted-suicide cases during the writing of this book), The Boys at Roy's (especially Ian and Peter), Dr. Julie Allen Whitehouse and "Professor" Pat Whitehouse (who were not formal roommates, but whose hearts were always open during the Clapham years, when I felt embraced as a member of the family), and Adam Heilman, Esq. (who was their Brooklyn equivalent).

Olga Sekulic, Esq., and Brian Derdowski, Esq., peppered me with questions every time we spoke or saw each other. Dr. Williamjames Hull Hoffer was generous with praise, gentle with encouraging guidance, and asked a lawyer's questions about my doctoral process, informed no doubt by his own doctoral experiences. In early stages, Jacquie Gauntlett, Margaret Savage (who, in a different time, would have been a professor herself), and Amor Vieira gave me pointed assistance, article clippings, and support beyond their administrative titles at the LSE. On several occasions throughout, Ed Guardaro Jr. treated me to friendly lunches and lectures, and more than once called me both for, and to give, advice related to this project. Dr. Molly Easo Smith, president of Manhattanville College, supported several of my initiatives during her tenure as dean of the Faculty of Arts and Sciences at Seton Hall University, and always encouraged me to finish my dissertation, ultimately ensuring I had the time and money to do so. One of these initiatives, a Faulty Innovation Grant under the Teaching, Learning and Technology Center, directly led to much of the analytical process and development regarding the role of the media in the Kevorkian cases, with technological and moral support from Drew Tatuskso, then senior instructional designer at Seton Hall University, and currently Program III director at Mount Aloysius College.

Two weeks before the viva of the original draft of the dissertation, I invited an interdisciplinary group to a pre-viva event to "sock it to me." Each member of the Sock It to Me Brain Trust read a field chapter and questioned me from the left, the right, and the center for the better part of four hours. This esteemed group included Olga Sekulic and Brian Derdowski (Chapter 3/chief prosecuting attorneys and judges in the Kevorkian cases and Chapter 4/juries in the Kevorkian cases), Dana Richardson (Chapter 5/families in the Kevorkian cases), Lisa Haddock and Nancy Palmstrom (Chapter 6/the role of the media in the Kevorkian cases) and Adam Heilman (Chapter 6). Honorary members Bob McGreevy (Chapter 3) and Rob Weinberg (Chapter 4) could not make the event, but graciously gave time and energy to lengthy phone "question times" from across a river and a country, respectively. All of these people presented difficult questions, complicated challenges, and artful insights, and surely made me more adept at speaking extemporaneously about what I had researched and analyzed. This unofficial group served as the link between my Thanksgiving Kitchen Committee (the Gasarches) and my official pre-viva meeting with my actual committee, composed of my supervisors, Professors Paul Rock and Robert Reiner. To all, I owe the mental agility that I brought into the viva and any courage that I displayed in defending the doctoral work to Professor David Downes and Professor Glennys Howarth.

In addition to the obvious readers who supervised me (Paul Rock, Robert Reiner, and Leonard Leigh) at the LSE, there were others who kindly read and commented. I particularly thank Robert P. McGreevy, Esq. (judges and prosecutors), Appellate Division Deputy Chief Clerk David Spokony (juries), Dr. Linda Wasserman (juries), and Barry T. Bassis, Esq. Although not readers of the dissertation, Professor Arthur Caplan and Dr. Steve Miles read a number of my early essays in the mid-1990s, as I started to delve into medically hastened death from a bioethics perspective, and Steve Miles kindly read a formative essay that ultimately was a cornerstone of my methodology discussion. Similarly, Sarah Colwell (who kindly supervised the original book proposal), Sandy Towers (who read articles and reports and enthusiastically cheered me on in the doctorate, as well as in this project), David Paige (who expressed belief in the project on crucial occasions, one of which was the day I was giving a medical humanities talk in Michigan regarding Chapters 5 and 7 of this work) and the incomparable Denver Compton (who was always available, constantly in communication, and nothing short of a Lamaze coach in the birthing of the final submission, while telling me that I was "going at [the final stages] like a champ," perhaps the only time in my life that anyone has paid me the compliment of referring to me as a boxer) at Greenwood Publishing all made insightful comments relating to my chapters as I set to the task of reconstructing historical debates of the medical euthanasia and physician-assisted suicide debate in the United States, with special reference to prosecutions of doctors, and some of the theoretical and legislative debates that related thereto. Ms. Erin Ryan was a great development editor (who answered loads of questions and gave background information that this neophyte book author found highly instructive). Ms. Michelle Scott was the gracious and informative ABC-CLIO project coordinator for this book; Mr. Sethu Baskaran Natarajan oversaw the copyediting, typesetting, proofreading, and indexing; Mr. Matthew Van Atta was the (gentle) copyeditor of this project (any remaining errors are solely my own). Ms. Pamela Jacobs, as editor-in-chief of the *New York Resident*, graciously allowed me to review films and documentaries about Jack Kevorkian and about the Oregon experience, and did so with interest and curiosity about the topic, as well as the reviewed films, all of which encouraged me.

The Tuesday meetings of the Mannheim Centre for the Study of Criminology and Criminal Justice, the Friday meetings of the Law Department Methodology Clinic (chaired by Mr. Peter Muchlinski), the Tuesday night meetings of the Sociology Department (chaired variously by Professor Leslie Sklair, Mr. Michael Burrage, and Professor Anthony Smith) were jointly and severally responsible for expanding my horizons, inviting me to give presentations as I struggled through the various aspects of working on evolving cases and legislation, and moving from adversarial to inquisitorial questions and interview styles. Dr. Lynne Jurgielwicz took me under her wing in the Friday meetings, and I am so grateful to her for encouraging me to do the undoable. Miss Angela White, the former departmental administrator of the Law Department at the LSE, and Mrs. Rachel Yarham, the doctoral program administrator at the LSE, assisted me with administrative information and assistance on occasions too numerous to count. In the Graduate School, Mr. Richard Leppington provided answers to questions at crucial moments, Ms. Sarah Johnson provided good news and excellent guidance, and Ms. Alexandra Williams communicated vitally important information. Miss Hannah Cocking in the Scholarships Office was an information resource who directly led to funding sources.

Students at Seton Hall University who provided cheerful technical assistance in photocopying and downloading during the late doctoral stages (of materials that all but the most technologically "unsavvy" could easily locate online), included Jeff DeCruz, Diana Giampicolo, and Megan Natale. These individuals were enormously entertained by my status as a techno-moron, and were kind enough to take my searches and produce reams of articles for me to review, saving me time and energy.

Without certain grants of money, with which I purchased equipment, travel tickets and accommodation, conference presentation opportunities, and that most precious asset of time, this work could never have been attempted. Thanks are due double or triple the amount of pounds and dollars I received from the United Kingdom Overseas Research Scheme, the LSE Morris Finer Memorial Fellowship, the LSE W. G. Hart Bursary in Law, and the Seton Hall University Faculty Innovation Grant of the Teaching, Learning and Technology Center. Particular field opportunities were funded by the University of London Irwin Trust, the National Long Term Resource Center of the University of Minnesota, the Center for Biomedical Ethics of the University of Minnesota, the Sociology Department and the Law Department of the LSE, and by faculty development funds from St. Francis College. Provost Timothy Houlihan, Academic Dean Michele Hirsch, and Professor Jaskiran Mathur of the Department of Sociology and Criminal Justice each deserve special mention; Dean Hirsch especially so for her comment after reading a doctoral chapter of mine, and remarking that I know things that nobody else knows and that I am so matter of fact (which made me evaluate anew, although that surely was not her intention in the compliment).

Libraries that kindly extended me privileges included that London School of Economics (which also provided my flatmate, Brendan), the Institute for Advanced Legal Studies, Kings College of London, New York University (Bobst and Law Libraries), Hamline University School of Law, Seton Hall University, Bloomfield College, and St. Francis College, as well as the New York City Bar Association. Richard Tuske, director of the Library at the New York City Bar Association, gave excellent information regarding my research efforts, on several occasions. Alex Kustanovich, reference librarian at St. Francis College, routinely gave my classes tech tool seminars at which he graciously answered as many of my research technique questions as those of students. In addition, I thank my friend Robert Gelber, the former law librarian at the Appellate Division in New York, although his role was social (bringing me on historical and art walks and group outings, which he now does professionally as a retired historian and librarian), rather than professional.

There is also a small cadre of people in New Jersey, where the final writing of this project took place, who ensured that my periods of self-incarceration were punctuated by the sort of social opportunities that remind us, even (or especially) during our most intense times of creativity and productivity, that we are human and have human connections. These include Linda Bloom, Elena Smith, and Steve Campanella, as well as the "Troy Towerettes"—Ms. Meredith McCutcheon, Ms. Pamela Miles, and Ms. Georgie Pickney. Also included in this group is Don Madson (whose tai chi classes gave me meditative time I have found in no other part of my life, although they were meant to simply be part of rehabilitation after knee surgery).

I have surely more people to thank for anything that I have accomplished in this work (indeed, the number in Michigan alone would exceed the number of pages in this book, with the result that most will continue to be thanked individually), and any success that may be associated with it; any errors or ambiguities are, of course, solely my own responsibility.

Introduction

This book is a treatment of the euthanasia and assisted-suicide debate of approximately 100 years leading up to the present, with an emphasis on the past 50 years. Nobody is without an opinion on this topic and few are without a personal experience (whether of a family member, spouse, friend, or colleague) that has not informed that opinion. Such personal experiences may lead inexorably to the conclusion that both the right to life and the right to die should be protected, or that earnest debate results in personal and professional ambivalence. During 1993–1994, the Michigan Commission on Death and Dying had two members who were eminent prosecutors generally and, insofar as this book is concerned, famous for prosecuting Jack "Dr. Death" Kevorkian specifically. John O'Hair was avidly pro-choice, while the other, Richard Thompson, was avidly pro-life.

During the same period, the House of Lords Select Committee on Medical Ethics was convening to consider similar issues, and the archbishop of York (who might be presumed to have a certain point of view) brought to the hearing and deliberations of evidence a past that included many family members in the medical profession, with a variety of points of view. The Baroness Flather, at the Debate of the Report of the House of Lords Select Committee on Medical Ethics (regarding medical euthanasia and the emergence of something called physician-assisted suicide), observed that members of the House of Lords variously approached her upon her appointment to the Select Committee. What surprised her was this:

More than one Member of your Lordships' House said to me, "Of course, you are totally pro-euthanasia, aren't you?" But as I walked down the corridor some other Member said to me, "I know that you are totally against euthanasia." I was delighted because they were as confused as I was. I had not cleared my thoughts on the subject.[1]

Perhaps one reason for this is that the debate begins with the language, the linguistics. The simple dictionary definition of euthanasia, offered by the *American Heritage College dic-tion-ary* (4th ed., p. 483), is that it is "the act or practice of ending the life of an individual suffering from a terminal illness or an incurable condition." On page 86, the same dictionary defines assisted suicide (a phrase that did not exist in a medical context prior to the 1990s) as "suicide accomplished with the aid of another person, especially a physician."

These are, however, incomplete definitions. As will be discussed, euthanasia may be active or passive, voluntary or non-voluntary, or involuntary. Active euthanasia refers to a positive act (for the most part by a physician), which is intended to bring about the death of the patient, such as the administration of lethal injection. Passive euthanasia generally denotes an act of commission or omission, whereby a patient's treatment, and/or nutrition and hydration, are discontinued. As a general matter, cases of double effect, in which painkillers prescribed for the purposes of palliation or relief of discomfort, but which have as a secondary effect the shortening of life, are not, as a general matter, considered to be euthanasia. Voluntary euthanasia is where a patient asks to have euthanasia administered (for example, a cancer patient who requests death). Non-voluntary euthanasia is where the act is performed without consent (as in cases where the patient is in a persistent vegetative state and thus incapable of giving consent). Involuntary euthanasia refers to situations where the "patient" neither consents nor wants to die (the Nazi atrocities serve as a powerful example of this). In much of the debate, mercy killing is distinguished from these acts, as nonmedical in intent or cause of death, usually with a humanitarian or compassionate motive (for example, the case of a man or woman who shoots, suffocates, or administers poison to a terminally ill spouse, friend, or child).

Matters of death and dying, and issues regarding euthanasia and what came to be known (during the 1990s) as assisted suicide, became a political hot topic during the last century. Particularly developed during the second half of the century, and most especially during the last decade of it, the debate has carried over into the twenty-first century in the form of legislative responses and prosecutorial policies toward doctors and other medical staffers, and also family members, who engage in hastening death. What used to be a private family matter has become one of public record in the criminal and civil courts, voter initiative and legislative task forces, medical care and legal liability, theological teachings, and sociological trends. While this book takes a somewhat comparative approach in terms of raising the pros and cons (and inviting the reader to draw his or her own conclusion), there is as much to be said for case-by-case debate and discourse as there is for full-blown regulation, legislation, and litigation.

A nonpartisan approach would be best in discussing, and perhaps in discovering, the debate. However, both proponents and opponents of hastened ending of life (an awkward, but all-embracing, neutral term) politicize the language of the euthanasia and assisted-suicide debates (plural deliberate), largely along pro-life and pro-choice lines. Hence, a disclaimer is necessary from the outset. There will be some linguistic shifts, particularly in quotations. A seeming inconsistency, this actually aspires to be consistent with the writers and speakers quoted. As one such example, the same decedent (another awkward, but all-embracing, neutral term) of a case in which Jack Kevorkian was prosecuted (indeed, almost any case in which Kevorkian was prosecuted during the 1990s) may be referred to as a "patient," a "victim," or a "client," depending upon whether a family member, prosecuting authority, or academic is discussing the matter, irrespective of "choice" issues (which becomes much more politicized when the debate is taken into account).

Pro-life advocates—whether their objections are religious or nonreligious in nature—contend that to phrase euthanasia and assisted suicide as a "choice" issue undercuts vulnerable populations (especially the disabled, the elderly, and racial or ethnic minorities), who will be at risk of having very little choice at all and be manipulated into hastening the end of life for economic and other reasons. Those of a religious background might argue that the sanctity of life is in God's hands alone, while being joined in the opinion by secular thinkers, who cite the dreaded slippery slope to the Nazi atrocities.

Pro-choice advocates contend that to phrase the issues as "pro-life" violates legal and ethical principles of self-determination, liberty, and equal treatment options. While, unlike abortion cases, euthanasia and assisted-suicide cases regard a person who has already been born, there are serious issues of capacity and competency. Indeed, this book focuses heavily upon those who are of age and mentally competent, and eschews discussion of "voluntary euthanasia" of certain groups, such as so-called "defective infants," who, by their infancy alone, have no voice by which to express so-called voluntariness. In addition, while suicide has been decriminalized in the United States and Anglo-American legal systems no longer view it as a crime, as a matter of Anglo-American theoretical law, a person cannot consent to his or her own homicide, which raises an insurmountable obstacle for those advocating consensual medical euthanasia.

These are only some of the thorny issues that arise and are vigorously debated, even as some states criminalize assisted suicide (like Michigan) and others decriminalize and regulate physician-assisted suicide (like Oregon and Washington). One question immediately arises: What led to this explosion of debate into "medical aid in dying," a term that euphemizes the concept of medical assistance in the termination of life, be it by euthanasia (as performed by members of the health care team) and physician-assisted suicide? It has arguably become the most important issue facing the medical profession (and a number of those professions having relation to medicine, including law, bioethics, and sociology, to name a few).

Axiomatic to developing a historical perspective of medical euthanasia is the definition of history as the "continuous methodical record of important or public events . . . the study of past events, especially of human affairs."[2] Thus, accepting that history is the record of human society, developing a historical perspective of medical euthanasia is inextricably intertwined with the fact that essays advocating active euthanasia in the context of modern medicine first appeared in the United States and England at the end of the 1800s, notwithstanding millennia of debate of the topic in other disciplines.

This should not be seen as too surprising, given that the modern hospital is considered to have existed for less than 150 years and that its function and efficacy have changed dramatically. Hoefler and Kamoie observed that before the turn of the twentieth century, there existed patient mortality rates of 25 percent and medical *staff* mortality rates of some 10 percent per year; many members of both groups succumbed to acute infections, derivate of their presence, rather than their role, in hospitals.[3]

With the defeat of acute disease, degenerative illnesses, also coming into prominence at the end of the nineteenth and the beginning of the twentieth centuries, fueled the beginnings of the modern medical euthanasia controversy. This increasing prominence of degenerative and late-onset illnesses is, of course, consistent with the societal changes, whereby life expectancy doubled from the norm of 40 years in 1851.[4]

From the 1930s onward, what had been a medical discussion of how to end patient suffering also became a political discussion and subject of debate. This is the focal point of the book. Done properly, a reader on either end of the pro-choice or pro-life spectrum will be vexed by a book declining to take a favorable or an unfavorable position regarding the issues, but instead follows the example of the Baroness Flather. The book is not a position statement; rather, it is an invitation to a broad readership across the spectrum of the debate to consider how different sides frame the debate. In this last regard, a reader should be constantly reminded that there are more than two sides to this debate, and in fact, there may be as many sides as there are people who consider it. As the debate developed, it went from being black and white to being shades of gray. Some of the cases reflect this, in pragmatic applications of the law, rather than per se applications of black-letter law or regulatory matters regarding health care providers, family members, and those who assess civil and criminal liability. This, too, is historic in nature, and gives context to what seems to be a perpetually ongoing reframing of the questions, as both personal matters and public issues.

Done properly, this book will ruffle feathers on both (or all) sides of the debate, yet seek a broad readership. To accomplish the latter, the narrative aims to develop how the debate began, how it developed to reach its present state, and to consider the way(s) forward for further discussion and debate. To do this, the narrative will have to provide facts for explanation as well as development of opposing arguments and emerging nuances. That said, for most of the history of modern medicine, euthanasia has been an illicit practice, without much in the way of empirical information or data about its practice. Thus, prior to the first trials in the 1950s, this book presents a historical debate presented by modern medicine in Chapter 1, starting with the late 1800s and into the early 1900s. Chapter 2 considers the good and the bad of the time period between the 1930s and the 1950s.

It is, however, Chapter 3 in which an illicit practice, debated in the hypothetical, becomes rooted in reality, in landmark cases that prescribe medical practice and provoke further debate and argument. On either side of the Atlantic Ocean, we see the prosecutions against medical practitioners and the emergence of the first major defenses to be asserted in regard to euthanasia. In the trial of New Hampshire's Hermann Sander, the first defense, regarding lack of proof of causation, emerges to the fore; while a jury instruction in the trial of John Bodkin Adams is instructive as to the defense of an unintended secondary effect of pain medication, which shortens life. These cases led to a flurry of debate and writing in the same decade by some of the greatest legal thinkers of the era, which provided the kindling for much of what would follow for the rest of the decade.

Chapter 4, which looks at seminal developments during the period from the 1960s through the 1980s, is where the debate moves from being the province of the professions to that of the "regular Joe." In England, suicide was officially decriminalized (although assisting in a suicide remained a serious felony), reflecting a sea change in attitudes that would develop 30 years later. Also, because of advances in medical technology, families (as well as medical professionals) found themselves confronting the questions that artificial life support raise, in cases that reached the highest courts. During the same period, the law and the American attitude was systematically destigmatizing suicide, from which the assisted-suicide debate of the 1990s would flow.

Chapter 5 considers a number of trials, many cases, and one doctor—Jack Kevorkian. Indeed, the various cases of *People v. Kevorkian* provided a number of landmark trials and a flurry of legislative activity, and it touched upon nearly every aspect of the euthanasia debate in modern medicine. In addition, the *Kevorkian* cases added a new phrase to the medical, legal, and general dictionaries—"assisted suicide." These thorny and overlapping cases of both euthanasia and assisted suicide, which went to the appellate courts of Michigan on more than one occasion, have a basic chronological arrangement, though at times the temporal arrangement of the trials trumps the events leading to the prosecutions. Both (or, more accurately, many) views of the debate are voiced by chief prosecuting attorneys, jurors, judges, and family members of the Kevorkian decedents.

Chapter 6 is, in a sense, a parallel text to Chapter 5, in that it examines legislative efforts and the wave of litigation brought by doctors, patients, and others during the 1990s, seeking to decriminalize and/or legalize medically assisted suicide. Cases on either side of the country found their way to the U.S. Supreme Court, which issued two rulings in 1997 in the companion cases of New York's *Quill v. Vacco* and *Washington v. Glucksberg*. These cases had, as their root, the unsuccessful prosecution of Dr. Timothy Quill, as well as concerns of fear of prosecution in Washington.

Chapter 7 regards the contemporaneously passed Oregon legislation that would allow doctors to assist in suicides, and a regulatory mechanism by which this could lawfully happen. This legislation continued to make news in the new millennium, in the case of *Gonzales v. Oregon*. That case regarded a ruling regarding whether the Oregon Death with Dignity Act was lawful or not (it was, but due to procedural grounds—a technical, rather than a substantive, decision). Many were disappointed that the Supreme Court declined to make a global statement about whether assisted suicide should be permitted or not, and instead it issued a ruling founded in administrative law.

Chapter 8 regards developments in euthanasia and assisted-suicide cases in the 2000s (as well as one case that was considered newsworthy as a right-to-die case, but rather was a protracted family dispute), Washington passed legislation allowing for assisted suicide, and there emerged a new method by which assisted-suicide laws could be applied—to family members who, in fact, engaged in mercy killing (as socially and legally constructed in a Georgia case). In addition, more case-by-case developments occurred, such as the exoneration of a physician who engaged in euthanasia in New Orleans in the wake of Hurricane Katrina. Last, the decade closed with a ruling by the Montana Supreme Court, providing for assisted suicide as a matter of state law.

Given the speed of new developments, these developments may themselves be historical in nature, and others may have emerged and become historical, even as this book goes to press. The debate continues to be one that those in favor of, and in opposition to, euthanasia upon request and assisted suicide argue with passion. That assisted suicide is legal in some states and criminal in others is a source of irritation to both sides. While many are opposed to euthanasia, the cases suggest that the law on the books and the law in practice differ widely, depending upon the factual scenario in any individual case.

While it is almost impossible to write a book on this topic in a neutral and nonpartisan way, when almost everyone debating the topic has strong and partisan views, this may be the only way to really examine the historical debates in context. Done well, a reader should not be able to discern my views; done very well, a reader should be able to see that compelling arguments are to be made by each side. Michael Ignatieff, writing the biographical *A Life: Isaiah Berlin*, put it well: "[s]ystems of values were never internally consistent. The conflict of values—liberty versus mercy; tolerance versus order; liberty versus social justice; resistance versus prudence—was intrinsic to human life."[5]

In the end, this book is an attempt to understand the euthanasia and assisted-suicide debates as they unfolded in the past, so as to be better prepared for the issues that unfurl in the present and in the future.

NOTES

1. Medical Ethics: Select Committee Report, HL Deb 09, May 1994, vol. 554, cc1344-412 (HMSO, 1994), http://hansard.millbanksystems.com/lords/1994/may/09/medical-ethics-select-committee-report, accessed February 25, 2011.
2. J. M. Hawkins, ed., *The Oxford Paperback Dictionary*, 3rd ed. (Oxford: Oxford University Press, 1990), 383.
3. J. M. Hoefler and B. E. Kamoie, *Deathright: Culture, Medicine, Politics and the Right to Die* (Colorado: Westview Press, 1994), 67.
4. S. Taylor, "Approaches to Health and Health Care," in S. Taylor and D. Field, eds., *Sociology of Health and Health Care: An Introduction for Nurses* (Oxford: Blackwell, 1993), 43–44.
5. Michael Ignatieff, *Isaiah Berlin: A Life* (New York: Metropolitan Books, 1998), 285.

1

Medical Euthanasia and Assisted Suicide: A Twentieth-Century Issue

INTRODUCTION

The birth of the modern American (medical) euthanasia controversy is placed around the year 1890,[1] coinciding (but not a coincidence) with the modernization of medicine. Thus, in considering the modern medical euthanasia controversy, a search for perspective commences in the early part of the twentieth century.

Sociologically, with its modernization, medicine has approached (and, some secular people might claim, replaced) religion as a major cultural and institutional molder. Indeed, Horace Miner's classical sociological/anthropological 1956 essay, "Body Ritual among the Nacirema," describing holy rituals among a quaint and strange holy man and holy home rituals is, in fact, a spoof of American (Nacirema spelled backward) medical culture, and how it is a national religion. In particular, Miner noted that "[t]he fundamental belief underlying the whole system appears that the human body is ugly and that its natural tendency is to debility and disease. Incarcerated in such a body, man's only hope is to avert these characteristics through the use of powerful influences of ritual and ceremony."[2] While Miner's intent was to comment humorously on the reliance of Americans on medical culture, he also effectively draws readers to the inevitable conclusion that modern medicine, like religion and the law, seeks to discover, control, and eradicate undesirable elements.[3]

Because of the breakthroughs of the early twentieth century, doctors were empowered to battle acute infections with success, and the hospital as an institution began its transformation from functioning essentially as a hospice to providing medical, surgical, and curative treatment for the sick and injured. For example, and indeed as the paradigmatic example, between 1886 and 1913, cancer deaths in New England and the Mid-Atlantic areas of the United States rose from 41 per 100,000 to 90 per 100,000. So it was that cancer became the first major referent for discussion of medical euthanasia and suicide in the early 1900s.

An early example of this, on the cusp of the twentieth century, was an 1899 editorial in the medical journal *The Lancet,* advising a physician to use morphine and chloroform to alleviate the pain of a patient with ovarian cancer. The piece read:

> We consider that a practitioner . . . perfectly justified in putting such treatment to an extreme degree, if that is the only way of affording freedom from acute suffering . . . [and] even should death result, the medical man has done the best he can for his patient.[4]

So it was that the medical profession began to promote a skewed use of the principle of double effect—that if the primary purpose of medication is to alleviate pain, the secondary effect of shortening life should not be deemed to be murder. In the 1950s, the legal profession, in the prosecution of Dr. John Bodkin Adams, would issue the first jury instruction charging the jury that if the purpose of medication was pain relief, and death was hastened as a result, criminal liability would not attach. As will be discussed, this was a very innovative way of defending against a criminal prosecution in the 1950s, and became a cornerstone of several of Jack "Dr. Death" Kevorkian's acquittals in the 1990s; yet the fact of the matter is that the medical profession itself was legitimating the practice at the turn of the century. As an aside, there are those who claim to be opposed to euthanasia (and assisted suicide), yet would say that they still have double effect.[5]

EARLY LEGALIZATION EFFORTS: OHIO

The first American proposal for legalizing medical euthanasia was made (and rejected) in Ohio in 1905. While the debate may not have emanated from the medical profession, medical advances served as the backdrop for the question of shortening life and hastening death. In Britain, for example, there is near universal notation of the "importance developments of analgesia and anesthesia."[6] Likewise, "as physicians who used the modern scientific method and modern principles of pharmacology consolidated their control over university and medical school training, the euthanasia debate entered the lay press and political forums [*sic*]."[7] It was in this climate that the Ohio legislature considered the question of whether to legalize euthanasia.

The Ohio legislature ultimately voted the bill down by a vote of 79–23, reflecting that a significant minority of the legislators actually favored the bill. The remarkable thing about this is that in 1905, there was a sea change showing that medical technology of the day had an effect on nearly a quarter of a cohort of legislators in their consideration of whether medical euthanasia should be permitted. Regardless of whether one favors or opposes medical euthanasia, the mere fact of this cultural shift is of import. That the bill was defeated is a historical fact, but that the bill was so seriously considered by so many is a historical perspective.

The bill provided that if one of "legal age and sound mind is fatally hurt, or so ill that recovery is impossible, is suffering great physical pain without hope of relief, his physician . . . may ask him or her in the presence of three witnesses, if he or she wants to be killed."[8] As an initial matter, the bill required an adult with decision-making capacity, although this early legislative effort did not enumerate particulars in this regard.

In this failed bill, there were requirements such as a terminal illness (as in Oregon and Washington). Alternatively permitted and predictive of degenerative illnesses that would plague longer-lived patients in the twentieth century (theoretically in cases such as Alzheimer's, Huntington's disease, ALS, MS, or end-stage AIDS, but with the caution that the sound-mind requirement would preclude the first two long before the terminal stages would be reached) were within the ambit of the bill proposed. Also in the conjunctive, the failed bill considered eligible those with unremitting pain (not permitted under Anglo-American law) or suffering (which cuts a wide swath of symptoms and varies from patient to patient).

The requirement of three witnesses, as a practical matter, may well have mitigated against coercion of the vulnerable; however, one possible criticism of this regards who the three witnesses are and whether that numerosity could have a coercive effect. Also in a trio, and in addition, if the patient did say they wanted to be killed, "three other physicians are to be called in, and if the agree that the case is hopeless, they are to proceed to do the job in a neat and convenient way with any anesthetic that may suit."[9]

The *New York Times* criticized the bill at the time of its second reading, under the auspices of Anna S. Hall, saying, "the efforts of this amazing representative of feminine gentleness and wisdom to legalize the murder of the aged, the sick and the injured have received the support of"[10] a physician "and a few other folks with unregulated enthusiasms."[11]

This language, largely ignored, served to predict two important future developments. First, there is a gendered commentary suggesting that women might be the prime movers of such a request (or might promote such a request). That may or may not have been accurate as to Ohio, but it certainly would prove to be so in Michigan during the 1990s Kevorkian cases. The second commentary is suggestive and critical of those with medical training as having a lack of regulation. As something of an irony, the prosecutions for medical euthanasia in the 1950s were due not to a lack of regulation of physicians, but *because* of such regulation of physicians. Further, in the 1990s Oregon legislation and the 2000s Washington legislation, there would be a litany of regulations and protocols, rather than the claimed "unregulated enthusiasms." Additionally, from a legal standpoint, the criminal law would, in 2005, provide for mitigation (indeed, exoneration) in the clearly unregulated case in post-Katrina New Orleans and mitigation of a Georgia mother who shot her two sons while they lay, side by side, in a nursing home, i.e., a medical facility. These whiffs of necessity, duress, and extreme emotional distress were in fact that unregulated (albeit not enthusiastic in the least).

However, the *Times* also contended that "the mitigating feature of this bill [was] its utter impracticability, for of course, no doctors above the grade of common hangmen would undertake such an atrocious and always completely unjustifiable task."[12] In this criticism, the unnamed staff (or editorial) writer predicted the coming of Yale Kamisar, whose writings in the 1950s and forward would reflect nonreligious (or secular) objections to euthanasia, and later, to his fellow Michigan resident Jack Kevorkian.

The defeat of this early bill, with its predictions of regulatory schemes to follow 100 years later, had one other noteworthy aspect. This local matter of a defeated bill on an obscure issue, previously unraised, captured the attention of a national newspaper. In a sense, that may have been predictive of the media frenzy that would accompany the criminal and civil litigation cases of the 1990s, the legislative efforts (both failed and successful) of the 1990s and 2000s, and the way in which the underlying medical issues would capture the imagination of the American people, many of whom either had family or friends or were themselves facing some of the health issues that the Ohio legislative effort initially sought to contemplate.

LEGISLATIVE EFFORTS REDUX: IOWA

During 1906, barely months after the Ohio legislative effort to legalize medical euthanasia failed, Iowa also considered a bill regarding euthanasia. On March 10, 1906, Representative F. N. Buckingham introduced a bill in the House entitled, "An Act to Require Physicians to Take Human Life under Certain Conditions."[13] This bill also required the consent of the patient, "suffering from an incurable disease,"[14] before the patient's life could be medically ended by use of an anesthetic. In addition to the three-doctor requirement that was similar to the Ohio bill, the proposed Iowa bill also required the recommendation of the coroner.

Two factors would have come into play under these provisions. First, and as would be debated at length in the coming decades, was the question of what is suffering and what is an incurable disease. As an example of the pernicious aspects of these questions, one of Jack Kevorkian's first cases, in 1991, for which he was tried and acquitted, was of a woman who had vulvadenia (euphemized as chronic pelvic pain in the press, this is actually a vaginal pain matter). Marjorie Wantz unquestionably had pain, had sleeplessness related thereto, but she also had depression and suicide attempts prior to her meeting with (and having her death hastened by) Kevorkian. Was she suffering? Certainly. Was her condition incurable? Evidently, given surgical and pain management attempts. But consider, as a sociological construct or a medical construct, whether Kevorkian would have participated in an assisted suicide of one with a male anatomy-specific complaint. (Indeed, during trial in 1996, and offered as evidence that he turned down people who did not have intractable suffering or an incurable illness, was that he refused to assist a man with knee issues and related pain with a hastened death, referring to him as "pathetic."[15]) While beyond the scope of this particular discussion, the question of whether this woman, who was depressed to the point of more than one suicide attempt, would have qualified retrospectively in Iowa, or prospectively under the regulatory schemes in Oregon and Washington, which have stringent psychological requirements.

Assemblyman Ross Hammond Gregory was also credited with introducing the Iowa bill in 1906, which has been characterized as "a measure far more extreme than the legislation proposed by Miss Hall."[16] The Ohio bill was designed to allow physicians to end patients' lives (or to assist in so doing), but the Iowa bill provided for hastened death of those over 10 years of age.[17] In almost every other sense, this would be deemed to be an infant incapable of, and unable to, consent and without capacity. Two family members within the state would have been required to be notified, or, in the alternative, the patient's guardian. This, too, raises a specter—if a patient needs a guardian to protect his or her own interests, then that points to a lack of legal capacity to consent *ab initio*. In such an event, even a patient-requested euthanasia would arguably be non-voluntary in nature, due to the lack of capacity to engage in a voluntary consent.

Assemblyman Gregory doubtless would have known, or had reason to know, this. He was a graduate of the Kentucky School of Medicine and had practiced surgery. Assemblyman Gregory's status as a professional doctor may be a reason why the *British Medical Journal* took an interest in the Iowa bill, and why it precipitated discussion in Britain as well. In its

March 17, 1906, issue, the journal published a short article and, according to N. D. A. Kemp, "with a short article in the *BMJ* . . . the question of euthanasia appear[ed] to have stirred people's imaginations."[18]

One of the concerns of the time was that the Iowa bill had parallels to the then-raging eugenics debate. Kemp argues that this is because the Iowa bill proposed that "persons suffering from hopeless disease or injury and hideously deformed or idiotic children should be put out of existence by the administration of an anesthetic."[19] Such a set of provisions would have gone beyond the parameters of voluntary euthanasia for competent adults with decision-making capacity into non-voluntary euthanasia of those who could not consent, and who were not of age. This is not what the social (let alone legal) construction of voluntary euthanasia is—if the patient cannot consent, engage in voluntariness, or has not capacity, then the euthanasia is, by definition not voluntary. (It may be voluntary on the part of the family member[s] or guardian[s] that requests it, but not on the part of the patient.)

The writers of the *BMJ* piece were reportedly not impressed by this, and even less so with the idea that a physician had been involved with the matter. The British medical establishment, by way of its journal, stated "[t]hat the man is either a crank or a particularly noxious type or a mere notoriety hunter, is clear enough from the statement attributed to him that he simply wish[ed] to make lawful that which is already practiced [*sic*] by the administration of an anesthetic."[20] This sort of excoriating peer commentary predicted the reception Jack Kevorkian would get in the 1990s from the American medical establishment, which rejected him in every possible way, including stripping him of his licenses to practice medicine in Michigan and California after his second and third (simultaneous) cases, of Marjorie Wantz and Sherry Miller, in 1991.

However, Representative (Dr.) Gregory was not perceived by his own Iowa community as a pillar of the community. He served more than one term (in stark contrast to Kevorkian, who resoundingly lost his bid for electoral office), had earned an excellent reputation as a physician, an "otherwise conservative track record who spent most of his time in the general assembly advancing the particular needs of the Corn Belt Meat Producers Association, although recognized as an 'influential member of the assembly,"[21] who had otherwise modest legislative goals, having reportedly only proposed one other piece of legislation, "to waive the examination requirement for veterinarians already registered in other states."[22] An additional sponsor of the bill, L. F. Summers was also a doctor, a wealthy businessman (owning interest in a pharmacy, a hardware store, and a farm), and an active member of the church—in short, he was a pillar of society. He might well have been viewed as a Renaissance man with medical leanings.

In any event, the legislation, sponsored even as it was by three prominent legislators, failed, as did the Ohio bill. They were said to have proposed it to begin the process of having the issue debated, with a long view toward passage years down the road. In a sense, they might have been early century midwestern analogues to New York's Dr. Timothy Quill, a respected hospice physician and medical school professor who was prosecuted briefly (but not indicted) in 1991, who then went on to sue the state of New York in federal court and to take his case, along with amici, to the U.S. Supreme Court. As with the trio of Iowa legislators, Quill did not prevail at the Supreme Court in 1997, but he did move the issue forward medically, legally, and in a positive light (he was considered the more proper counterpart, the better answer to, Jack Kevorkian, who was viewed ultimately as the poster boy for all that was wrong with the debate).

As with the Ohio bill, the Iowa bill received media attention as well as the medical journal attention noted earlier. One perspective, predictive of what would happen at the end of the same century, is that the legislative debate created the news that stimulated public imagination and contemplation. At the beginning of the century, this was promoted in large part because of new and improved benefits of analgesics; while at the end of the century, this was prompted by a group of populations who were afraid of degenerative illnesses or of being kept alive by machines as with Karen Ann Quinlan in the 1970s and Nancy Cruzan in the 1980s.

What is also of interest is that even at this early juncture in the century, it was widely accepted that there was a quiet underground practice of medical euthanasia. The early 1900s debates did not focus so much on religion or on morality as they did on practical matters, including reference to the Ohio law as the "chloroform" bill.

THE CRIME OF SUICIDE AND THE NON-CONCEPT OF ASSISTED SUICIDE

Absent from this early dialogue is the discussion of medically assisted suicide. This narrative seeks both to retrospectively explain that seeming legal lacunae, with a brief discussion of the crime of suicide, and to prospectively explain why assisting in a suicide was considered criminal, even when suicide itself ceased to be.

Contrary to popular belief, suicide was a crime in Anglo-American jurisdictions. In keeping with the ecclesiastical laws that prohibited suicidants from being buried on consecrated ground, because they were murderers who engaged in self-murder (i.e., suicide), under Anglo-American law suicidants were subject to civil penalties. This begs the question of how the state could effectively punish a person who commits suicide, and is presumably dead.

The answer to that question is that in England, suicidants were buried at cross-roads, their property was forfeited to the state, and their families were left without inheritance (and also left with social as well as legal stigma). What would today be viewed as immediately problematic within the Anglo-American theoretical and comparative criminal law is that the living family had state punishment imposed upon survivors, who presumably did not know about or intend for the death of the suicidant and did not (by definition of suicide as self-murder) cause the death by self-murder. In other words, innocent people were being punished for a crime for which they had no part in the commission.

It was only in the course of the decriminalization of suicide that these penalties were removed. However, that was not to say that the state was (or the states were) encouraging people to commit suicide for any reason (leaving aside altruistic suicide and military suicide missions). Indeed, even in the 2000s, we refer to someone who has taken his or her own life as having committed suicide. This linguistic construction is not minor or idiomatic; rather, it remains derivative of the phrase to commit homicide, the worst of all crimes (at least in terms of general sentencing and punishment matters). Furthermore, as a near universal in Anglo-American jurisdictions, a person who states that they are going to commit suicide is subject to civil commitment, as representing a harm to themselves and/or others. (As an aside, and deferred for discussion in the Kevorkian cases, one of the women for whose assistance in suicide he was tried, to acquittal, had attempted suicide and had a psychiatric history related thereto, of which the prosecutor made much at trial in arguing that Kevorkian should be convicted for preying upon a vulnerable patient, albeit unsuccessfully so as the verdict demonstrated.)

What was particularly interesting about the emerging medical euthanasia debate of the early 1900s was that none of the discussion revolved around a patient committing suicide; rather, and perhaps predictively of Kevorkian (viewed in either a positive or a negative light), the focus was upon physician control. That there was not a discussion about decriminalizing suicide for those who were terminally ill or in intractable pain speaks as much as anything to the way in which the debate was being shaped. After all, and to be discussed in a later chapter, suicide was a crime in England until 1961; various states decriminalized suicide at different times and for different reasons, but discussion of decriminalizing assisting in a suicide was not a developing legal issue until the 1990s. Indeed, in Kevorkian's home state of Michigan, there was an assisted-suicide prosecution and appeal in 1984, which will be discussed in that chapter. Thus, his proclamation that what he was doing could not be a crime, since suicide was not a crime (and, of course, conduct of medical euthanasia has never been decriminalized in Michigan or any state), was lacking in historical perspective.

The context, the perspective, of this brief narrative, is to offer a reader uninitiated in criminal law or legal theory an explanation and an invitation. As explanation is the fact that suicide was still a crime at the beginning of the twentieth century (and for some time thereafter in most Anglo-American jurisdictions). The invitation is to consider, to question, that the medical euthanasia debate began to emerge at the beginning of the century, whereas the physician-assisted suicide debate did not do so until the century's final decade.

CONCLUSION

Ohio and Iowa set the tone for more than just their own states. The Iowa trio was correct that it would take years before medical euthanasia would be contemplated again. However, it would not be the 20 or so years they had thought, The next state to consider euthanasia legislation, Nebraska, would not do so until 1937. It would be then that Senator John Comstock introduced a Voluntary Euthanasia Act in the U.S. Senate—one year before the National Society for the Legalization of Euthanasia would be founded, which soon became the Euthanasia Society of America. During the "quiet time," medicine continued to advance, and life spans continued to lengthen, leading to more death by degenerative illness (primarily viewed through the paradigm of cancer). The next chapter includes a discussion of how the debate regarding medical euthanasia emerged, even as the euthanasia of the King was administered and hidden from public knowledge.

NOTES

1. Kuepper, Stephen L., "Euthanasia in America, 1890–1960: The Controversy, the Movement and the Law" (PhD diss., Rutgers University, the State University of New Jersey at New Brunswick, 1981), 56n84.

2. Miner, Horace, "Body Ritual among the Nacirema," *American Anthropologist* 58, no. 3 (1956): 503–4.

3. Kearl, Michael C., *Endings: A Sociology of Death and Dying* (Oxford: Oxford University Press, 1989), 406.

4. Emanuel, Ezekiel J., "The History of the Euthanasia Debates in the United States and Britain," *Annals of Internal Medicine* 121, no. 10 (November 15, 1994): 793–802.

5. One such legislator, Michigan representative Nick Ciamaritaro, a pro-life advocate who passionately opposed assisted suicide and euthanasia when it was

being debated in Michigan in the 1990s as a result of Kevorkian's activities, said this in an interview on March 2, 1994.

6. Fye, W. Bruce, "Active Euthanasia: An Historical Survey of Its Conceptual Origins and Introduction into Medical Thought," *Bulletin of the History of Medicine* (1968), 493–502, cited in N. D. A. Kemp, *"Merciful Release": The History of the British Euthanasia Movement* (Manchester: Manchester University Press, 2002).

7. Manning, Michael, *Euthanasia and Physician-Assisted Suicide: Killing or Caring?* (Mahwah, NJ: Paulist Press, 1998).

8. "Topics of the Times," *New York Times*, January 25, 1906, ProQuest Historical Newspapers, *New York Times*, p. 8.

9. *Id.*

10. *Id.*

11. *Id.*

12. *Id.*

13. "Another Euthanasia Bill," Special to the *New York Times*, March 11, 1906, ProQuest Historical Newspapers, *New York Times*, p. 1.

14. *Id.*

15. Personal observation during trial: As an unlikely coincidence, I had put off surgery to repair a traumatic knee injury, specifically so as to be able to attend the two 1996 Kevorkian trials, and was using a metal brace and cane to walk, although other people seemed to recall that I was in fact using crutches, which does not comport with my recollection. However, after the question about the male would-be patient with knee pain came up in court, friends among the court watchers asked me about my knee experiences, which I described as being "chronically sore" during the preoperative stage. The purpose of this footnote is not to say that George Miller did not have pain, or that I was stoic, but rather to say that pain is perceived differently by different people at different times and under different circumstances, as is suffering, also a subjective measure.

16. Appel, Jacob M., "A Duty to Kill? A Duty to Die? Rethinking the Euthanasia Controversy of 1906," *Bulletin of the History of Medicine* 78, no. 3 (2004): 610–34.

17. *Id.*

18. Kemp, N. D. A., *"Merciful Release": The History of the British Euthanasia Movement*, 54.

19. *Id.*

20. *Id.*

21. Appel, *op. cit.*

22. *Id.*

2

The 1930s and 1940s: From the King of England to the Holocaust

INTRODUCTION

During the first half of the twentieth century, the 1930s and 1940s, and especially the Nazi euthanasia program, served to inform the assisted-suicide and euthanasia debate that would develop by the end of the century. In a very real sense, this time period is the least controversial of the debate, because it ultimately proved to be the most monstrous example of the slippery slope, whether viewed from a religious or a secular perspective. Less well known is that there were debates in Anglo-American jurisdictions, which also deserve discussion.

The two have in common that with the modernization of medicine, the medical institution began to compete mightily with religion as a social institution. That is to say, cultural fears regarding death, and desires for immortality, began to look to medical men, rather than men of God, for answers to the questions of living longer. As with religion and the law, modern medicine sought (and continues to seek) to discover, control, and eradicate undesirable elements.[1] The previous century provided breakthroughs that had invested in doctors' nearly miraculous powers to battle acute infection successfully, precipitating a geographical and social change of the hospital from de facto hospice to an institution that provided medical, surgical, and curative treatments for the sick and injured. Accordingly, and taken as axiomatic, the social construction of health and medicine changed.

Further arming the medical profession in the war against the devils of illness and the demons of death were technological advancements, which created a class of knights from previously mere-mortal medical professionals.

These technological advancements essentially empowered doctors to slay some of the dragons that would previously have brought death, thus allowing for new medical successes against untimely deaths and failing health. As a result of all of this, the role of doctors in relation to their dying patients began to look quite different, even more dissimilar than the roles of hospitals did.

Thus, with life spans increasing, along with the onset of degenerative illness, the medical profession began to respond. By the end of the period commencing with the 1930s and into the 1940s, the taking of lives with disabilities, and non-voluntary and involuntary euthanasia, were scandals effectively summarized with the words "Nazi atrocities." Neither the initial Anglo-American proponents nor the early German proponents of medical euthanasia intended for medical euthanasia to be available to those other than the elite.

THE DEATH OF KING GEORGE V

If doctors of the past were left to watch helplessly as their patients' bodies were overtaken by degenerative illness (in cases where opportunistic infection, such as pneumonia, the old man's friend, did not prevail), then modern physicians and medicine, with their treatments and cures, arrived in the 1900s to find a more interactive relationship between themselves and their clientele. By using this descriptive, the discussion of the debate includes immediate family and friends, as well as the patient, in the relationship. The experience of the King of England, and subsequent debate in the United Kingdom, shows the beginning of the trend toward medical euthanasia and/or allowing for the hastening of death, as an option of the patient and/or family.

From an academic stance, it was said that the personal physician of King George V, Lord Dawson of Penn, engaged in euthanasia of the King, after having ascertained that Queen Mary "saw no virtue in allowing the King's suffering to continue" and took whatever steps were needed to cut his suffering short.[2] In and of itself, the fact that a wife might want to see her husband's suffering end was perhaps not so significant; however, that it was the Queen asking for the King to be euthanized was. Proof of this is the fact that there was no public disclosure of the Queen's request, and Lord Dawson of Penn's notes were kept secret for 50 years, until 1986. It was only then that it came to light that "as he lay comatose on his deathbed in 1946, King George V was injected with fatal doses of morphine and cocaine to assure him a painless death in time, according to his physician's notes."[3]

As an initial matter, it should be noted that in slightly different terms, requesting the death of a monarch, without any other extenuating facts (or perhaps even with them), is technically nothing short of a technical attempt (even if sympathetically motivated by a compassionate motivation, where euthanasia was unlawful) of a coup d'état, with its result of a new monarch being installed. Moreover, from a technical legal standpoint from the perspective of the criminal law, Lord Dawson of Penn, by complying with such a request, technically assassinated the King, by being under hire (albeit as his personal physician) to hasten his death. This should be contrasted with the hastened death of another George, President George Washington (who actually died in retirement, not while in office).

The first former president developed a sore throat during 1799; during the subsequent two days before his death, and consistent with medical practices of the time, his physicians bled some five pints of blood (i.e., drain blood from his veins, a standard practice of the time, although now known to further weaken the system, if not to kill outright from medical exsanguination) from him. "While Washington's death was probably (and obviously inadvertently) as likely to have been caused by shock and loss of blood as by strep throat (for which he also gargled),"[4] it is undisputed that his doctors were seeking to save his life and to cure him, rather than to end his life (for any compassionate motivation) and to kill him. Ultimately, Washington's death may have been the first (albeit inadvertent) publicized physician-assisted suicide in the United States. Were he to have fallen ill in contemporary times, there is no doubt that as part of even the most basic form of medical care, "Washington would have received any number of antibiotics, retained his life blood and likely lived to die of one of the degenerative illnesses now common to old age."[5]

The death of King George V on January 20, 1936, by two injections administered by Lord Dawson, the royal physician, who recorded his conduct in notes, was anything but state of the art, and was indeed and unquestionably outside of the ambit of lawful conduct. In fact, the two injections were barely an hour and a half after Lord Dawson had written a brief medical bulletin to the effect that "The King's life is moving peacefully toward its close."[6] Reports at the time were (medically) to the effect that the King had a peaceful death at midnight. In other words, there was nothing other than that the 71-year-old King, who had been in failing or declining health as a result of bronchial distress for a number of months, which accelerated in the monarch's final four days, died peacefully (and naturally) in the night.

In what might have predicted the 1990s conduct of Dr. Nigel Cox, to be discussed in Chapter 6, the medical protocol of note taking and documentation was how the euthanasia of the King first came to light. This was years after Lord Dawson's biographer, Francis Watson, initially omitted reference to the King's euthanasia at Sandringham Castle, when Watson first prepared a biography in 1950, after Lord Dawson's death. The omission of this important piece of history had been at the request of Lord Dawson's widow, and Watson first disclosed this important, albeit ancillary, piece of information in 1986, with the regret that he had not included it in the original book. However, Watson did correct the record, not only with interviews with the press, but also in an article for the academic journal *History Today*, in a 1986 article entitled "The Death of George V."[7]

Watson himself learned about the physician's role in the King's death while researching a biography of Lord Dawson, whose notes were preserved in the Windsor Castle archives. Only after the physician's death, in preparation for the biography that was released in 1950, did the notes come to light. What might be viewed as a private matter in a family of ordinary people rose to the level of a conspiracy as to the King's euthanasia. In fact, according to the Dawson notes, he "had already taken the precaution of phoning his wife in London to ask that she 'advise the Times to hold back publication.' "[8] This showed a level of planning no less than that of the Kevorkian cases, at the end of the century, in which family members and physician joined—however, there is a clear exception here in that the patient, the King, was not the person issuing the request, but rather that the family was, with which the physician complied. Even in 1950, over a decade after the King's death and five years after Dawson's death, the conspiracy of silence continued, when Lady Dawson asked Watson to omit reference to the King's euthanasia, a request with which Watson complied.

It was only when the 79-year-old Watson was in the winter of his own life that he corrected the record, initially by authoring the academic (and potentially obscure) article. Watson wrote that the:

> Prince of Wales had earlier told Lord Dawson that he and the Queen had no wish for the King's life to be prolonged if the illness were judged to be mortal; but . . . left the decision in the doctor's hands.[9]

In Dawson's notes, as quoted by Watson, "[a]t about 11 o'clock it was evident that the last stage might endure for many hours, unknown to the Patient but little comporting with that dignity and serenity which he so richly merited and which demanded a brief final scene."[10] In words that almost predicted the sort of discussion that would emerge in the 1970s in the case of how to handle Karen Ann Quinlan's persistent vegetative state, and the 1980s matter of Nancy Cruzan, but without any of the attendant litigation, Dawson came to his own conclusions and decisions. He wrote that "[h]ours of waiting just for the mechanical end when all that is really life has departed only exhausts the onlookers and keeps them so strained that they cannot avail themselves of the solace of thought, communion or prayer."[11] Then, with a physician's sense of protocol and record, Dawson wrote that he "therefore decided to determine the end and injected (myself) morphia gr. 2/4 & [*sic*] shortly afterwards cocaine gr.1 into the distended jugular vein."[12] This physician's sense of protocol would, in 1991, find its way into the English Crown Court of Winchester, in the case against Nigel Cox, who recorded that he gave dying patient Lillian Boyes two ampoules of potassium chloride to stop the heart, for which he was prosecuted, convicted (only of attempted murder, because the cause of death could not be established), and given a suspended sentence, a reflection of the social reluctance to interfere with a private matter, which acknowledging the medical and legal necessity of punishment.

Dawson then noted that "the Queen and family returned and stood around the bedside—the Queen dignified and controlled others with tears, gentle but not noisy . . . life passed so quietly and gently that it would be difficult to determine the exact moment."[13] In a very real sense, this scenario makes the royal family part of, and party to, the euthanasia of the King. With an abundance of politeness and discretion, this was shown, perhaps for the first time, in the 2010 Oscar-winning film *The King's Speech*. Taken together with Watson's unreported (though documented) information about Lord Dawson of Penn's role in the King's medical euthanasia, finally disclosed 50 years after the fact, one might say that the King's death in a post-millennium world might have been reported and discussed, and fodder for debate, as indeed these retrospective reports and media depictions hint at.

There is another matter that, looking back, gives rise to controversy in regard to these events, albeit controversy that has gone largely undebated. Watson observed that the moment of death was recorded "as 11:55 pm, and the news was broadcast by the BBC at 12:10. . . . [O]ne of Dawson's considerations had been 'the importance of the death receiving its first announcement in the morning papers rather than the less appropriate evening journals.' "[14] In a different context, this might be analogized to the usual time of execution of Death Row inmates in the United States, although here, the timing of death and press was deemed to have a "compassionate" motivation. Analytically, objectors to euthanasia generally might consider this press seeking timing in juxtaposition to that of Michigan's doctor Jack "Dr. Death" Kevorkian in the 1990s (whose press-seeking efforts were unquestionably viewed in a negative light, as opposed to Dawson's genteel media manipulation, which was represented as gentlemanly discretion).

ANGLO-AMERICAN LEGISLATIVE EFFORTS IN THE 1930s AS PREDICTING LATE-CENTURY DEBATES

Prior to the 1930s (and indeed, until the 1950s), no English or American doctor was prosecuted for euthanasia or mercy killing. That does not, however, mean that the matter was not being pushed forward by doctors into the public eye and political ear. If the King's death was effected in the shadow of the law, privately attended and publicly unreported (in terms of the euthanasia aspect, at least for the then-present and the next 50 years), the theoretical doctor's role was being debated in the House of Lords, and among political action groups being formed for this very purpose on both sides of the Atlantic. In all likelihood, this was because the practice of medicine itself was changing. In addition, there was the openly accepted fact that while the law on the books may be written in stone (or at least in ink), the law in action was, to use Yale Kamisar's word, more "malleable." Kamisar, observing primarily family mercy killings prior to the 1950s, noted "the high incidence of failures to indict, acquittals, suspended sentences and reprieves."[15] Kamisar argued that these all led "considerable support to the view that if the circumstances are so compelling that the defendant ought to violate the law, then they are compelling enough for the jury to violate their oaths."[16]

However, the circumstances in the 1930s, Lord Dawson of Penn aside, regarded no legally certain waters for doctors, with their rising tide of cancer patients (especially) and other long-term illnesses. Pain and suffering were becoming dragons to do battle with, as patients began to live longer and die less frequently (and at earlier ages) of acute infections, accidents, and war. Moreover, while there were no doctor prosecutions yet in the 1930s, as Glanville Williams was to observe two decades later (and with the opportunity to review the first doctor prosecutions in the United States of Herman Sander, and the prosecution of England's Dr. John Bodkin Adams, mysteriously omitted from discussions among Kamisar, Williams, and Catholic legal theorist Norman St. John-Stevas), "the prospect of a sentimental acquittal cannot be reckoned as a certainty."[17] Moreover, a so-called sentimental acquittal (or jury nullification) was more likely in the case of a family member who engaged in mercy killing than a doctor who engaged in medical euthanasia, which, while similar in causation, would emanate from a different motive. Further, under the criminal law, intent (rather than motive) is the requisite mental state and element for conviction, and jury nullification was arguably the application of motive rather than intent to the facts at hand. Thus, a jury might be more inclined to engage in a "sentimental acquittal" in a case where sentimentality was more likely to be evoked—a husband, a wife, a child, a parent, a loved one, not so much a patient, a client, more likely to be considered to be a victim of crime, rather than treatment of medicine.

The Voluntary Euthanasia Legalisation Society in England was founded in 1935,[18] and its American counterpart, the Euthanasia Society of America, formed in 1948.[19] These may have shared a root in comments by Dr. Charles Killick Millard in a 1931 presidential address to an annual meeting of the Society of Medical Officers of Health. In these remarks, "Millard advocated the 'Legalisation of Voluntary Euthanasia [and] provoked the first, sustained, nationwide debate [in England] . . . [in] leading medical journals, daily newspapers and popular periodicals, eliciting a broad spectrum of opinion from among the medical profession, ecclesiastics and lay persons."[20]

The House of Lords ultimately debated a bill to legalize voluntary euthanasia in December 1936. This debate, nearly a year after the King's January 1936 death, his euthanasia at the hand of Lord Dawson of Penn, had—as a voice of opposition to legalization—Lord Dawson of Penn. While in general sympathy of the bill, the unadmitted euthanasia provider "scouted the opposition notion that the only duty of the doctor is to save life. That, he said, was the nineteenth-century view, but medical opinion has changed, we now think it is also the doctor's duty to relieve pain."[21] Citing the advantages of anesthesia in childbirth and dentistry, painkillers and sleep medication were something that he focused upon, saying there was "a different sense of values from the ages which have gone before. It looks upon life more from the quality than of the quantity. It places less value on life when its usefulness has come to an end."[22]

Consider the source—a medical doctor, indeed the same medical doctor who had surreptitiously ended the King's life. Whether compassionate or killer, the fact is that Lord Dawson was also in the role of legislator, a unique role and dual perspective of which likely no other doctor has been in the position. But for the courtesies of being the King's physician and acting at the Queen's behest, in effect a medical murderer was being permitted to debate and discuss the possibilities of changing the law regarding medical murder. For all of the pro-life and pro-choice views that might be espoused, this searing conflict of interest has gone without discussion. When the legal philosophers of the 1950s were writing, it was not known, and in time since, it has simply gone largely unaddressed.

What has not gone without comment is that the prime movers in the 1930s were doctors—doctors who were working in a medical world changed by new drugs and treatments as friends, with degenerative illnesses beginning to become prominent. Thus, a few of the concepts, which would come to be relevant and ruling at the end of the twentieth century, began to emerge as well.

For example, the 1936 House of Lords debated who would be eligible for euthanasia. The list is not unlike that of Oregon in the 1990s and Washington State in the 2000s. In order to be eligible, the English version of the bill required an "eligible patient to be 21 or over [that is to say, a person of majority, capable of legal consent and not an infant], and 'suffering from a disease involving severe pain and of an incurable and fatal character,' to forward a specially prescribed application."[23] In addition, the bill required "two medical certificates, one signed by the attending physician, and the other by a specially qualified physician . . . a personal interview with the patient . . . and that the patient fully understands the nature and purpose of the application."[24]

Taking off from the King's death, Lord Dawson himself voted in opposition to the Voluntary Euthanasia (Legislation) Bill in the House of Lords on December 1, 1936. In his vote against a second reading of the bill, he argued that euthanasia was "something which belongs to the wisdom and conscience of the medical profession and not to the realm of the law. . . . The machinery of this Bill would turn the sick room into a bureau and be destructive of [physicians'] usefulness."[25] Lord Dawson went on to say that "I believe not only that the law would remain nugatory but that it would deter those who are, as I think, carrying out the mission of mercy."[26]

This view that any euthanasia bill would make life more (rather than less) difficult for Anglo-American doctors was not solely held by the royal euthanasist. Dr. Harry Roberts, while in accord with the general aims of the Euthanasia Society, also protested that limitations and conditions (which in post-millennial parlance would be called procedural safeguards, in states where physician-assisted suicide is lawful), were so narrow in their application, that legal medical euthanasia would be of no avail to much of the medical practices that were being conducted within the shadows of the law. Writing in *Euthanasia and Other Aspects of Life and Death*, in 1936, he wrote that:

> When a good physician can keep life no longer in, he makes a fair and easy passage for it to be out. . . . [S]uch I think would be regarded as the bounded duty (whatever the law may say) of every practicing [*sic*] doctor).[27]

Roberts continued on, taking personal, as well as professional, responsibility for conduct outside the ambit of the law.

> I would not hesitate painlessly to end the life . . . regardless of the convention or formal legality. When my sympathy outweighs my fear of, and respect for, the law, I obey the orders of the former.[28]

In other words, some who opposed legalized euthanasia did so on the grounds that physicians would be too constrained in what they were permitted to do that doctors would—to use a phrase that would ring in the iconic Kevorkian landmark narrative interview to *60 Minutes* in 1998, and his jury conviction in 1999—not have "more control." Thus, while religious opposition, such as that of the Catholic Church, might have been predicted,[29] part of the reason for the vote of 35–14 in the House of Lords pertained to doctors wanting *more*, not less, in terms of permissible euthanasia. Some doctors, such as Dr. Kenneth McFadyean, viewed any legal safeguards and protocols as intrusive to the practice of medicine. McFadyean, commenting on the English bill, observed "from a public platform that he had practiced euthanasia for 20 years and he did not believe he was running risks because he helped a hopeless sufferer out of this life."[30]

Thus, while some viewed administering medical euthanasia as violating their Hippocratic Oath to "give no deadly medicine to any one if asked, nor suggest such counsel,"[31] others apparently viewed the physician's role as shifting. However, whereas some doctors viewed the changing medical technology as placing more patients in the position of seeking euthanasia, others took the view that advances in what would become the palliative care movement were to the contrary. This was noted in the House of Lords debates, where none other than Lord Dawson of Penn observed that "in less than half of the cases of fatal cancer [the paradigmatic referent in the 1930s] . . . pain is an outstanding feature."[32]

What was this "more"? In 1938, the American Society originally "had at first intended to include compulsory euthanasia for monstrosities and imbeciles in its programme."[33] Thus, in 1937, Nebraska senator John Comstock introduced legislation entitled the Voluntary Euthanasia Act, which went without vote or approval but showed the American interest in the subject.[34] A similar effort to introduce a law took place with a bill in 1930s New York State (which would provide constitutional challenge in the 1990s, when Dr. Timothy Quill would give a cancer patient a prescription for a lethal overdose of barbiturates also failed).

As something of an irony, there is a "more" where doctors have a certain level of free reign, as long as they are discreet in the endeavor—that of the double effect. The principle of double effect is that if the primary purpose of a drug is to alleviate pain and/or suffering, and it secondarily shortens the life of the patient, there is no legal intent to commit a crime, even if a patient is medically caused to die sooner. As if foreseeing the 1950s trial of England's Dr. John Bodkin Adams, the first in which the judicial instruction of double effect was issued to a jury to consider in its finding of facts, Lord Dawson of Penn argued in his speech to the House of Lords that the "distinction between injecting a drug with the express intention of killing the patient, and injecting the same drug in order to ease his pain, in the knowledge that his expectation of life would be greatly reduced," was generally accepted within the medical profession.[35]

What, as Yale Kamisar noted, esteemed physician Lord Horder commented in the House of Lords debates that, "during the morning depression he [the patient] will be found to favour . . . [euthanasia] . . . later in the day he will think quite differently, or will have forgotten all about it . . . [with] counterpart in the alternating moods and confused judgments of the sick."[36] Lord Horder's concerns (reaffirmed in 1950) would also find their way into later assisted-suicide legislation—in Oregon and Washington State, any repudiation or vacillation is deemed a recantation, and in fact should start the request clock again.

The converse of the differences in patient mood is the possible differences in doctor diagnosis. To some degree, the English and American proposals in the 1930s mitigated against this by requiring more than one doctor to be involved in the decision-making process. However, as noted above, doctors themselves chafed at this intrusion on the sanctity of their own doctor-patient relationships. A previously ardent supporter of the euthanasia movement generally (and, ironically, more for those who were insane or under a mental defect, and thus not within the ambit of the voluntary euthanasia bills proposed in England and the United States), Dr. Abraham L. Wolbarst, was cited by Kamisar for his recognition of "the difficulty involved in the decision as to incurability" as cause to reconsider, because "doctors are only human beings with few if any supermen among them. They make honest mistakes, like other men, because of the limitations of the human mind."[37] As an aside, one of the mistakes did not regard illness of the patient, but rather regarded the suffering of the family members; or as St. John-Stevas put it, "the question must also be asked, whose suffering is to be alleviated, that of the patient or that of the relatives? The scope for self-deception is considerable and an apparently humanitarian motive may be only a cloak for selfishness."[38] Thus, doctors were subject to the challenges of medical mistake (as differentiated from failure to anticipate new and innovative treatments and cures), and also to the challenge of the source of the request for euthanasia, since a family request is (at best) a reflection of non-voluntary (or potentially involuntary) medical euthanasia of a patient.

THE RELATIONSHIP OF THE NAZI ATROCITIES UPON THE EUTHANASIA DEBATE

Of course, as the late 1930s and the first half of the 1940s would show, not all doctors were making mere "mistakes" in diagnoses, nor was patient voluntariness (or consent) a cultural universal. While the English Voluntary Euthanasia Society and the Euthanasia Society of America were limiting their efforts to legalize euthanasia to cases of voluntary terminally ill adults with capacity (rather than the mentally ill and disabled infants without capacity), the Germans were moving in the opposite direction.

Courts making precedent-setting decisions, legislative committees making recommendations to governments, and those who are making medical policy look with trepidation at the historical lessons of the Nazi atrocities. Although, as Kamisar noted, the "small beginnings: of the German euthanasia program were seemingly benign"[39] with German Jews excluded from the program because "the blessing of euthanasia should only be granted to [true] Germans," ultimately 275,000 people were involuntarily exterminated in the so-called "euthanasia centers."[40] Euphemized as euthanasia, the Nazi experience was in fact one of targeted genocide—toward Jews, homosexuals, gypsies, and those who were mentally and/or physically handicapped. There is no question that one of the greatest fears concerning the euthanasia and assisted-suicide debate in the second half of the twentieth century (and ongoing into the twenty-first century) for decriminalization and legalization derived (and still derives) from the Nazi euthanasia program; nor is there any dispute that no historical event has had the worldwide implications that the Nazi era did. It was a time during which the slippery slope went from a theoretical to a practical example (in both senses of the phrase, during which euthanasia went from a discussion of voluntary (albeit primarily on behalf of others) to non-voluntary, to the most grotesquely involuntary euthanasia program imaginable. The Nazi era was characterized not by the use of medical technology, but rather of abuses of it, along with the systematic erosion of medical ethics.

As it happens, the Nazi euthanasia program has certain parallels to (as well as glaring differences from) the English experience, although the former not at first blush. For example, the first full-scale study of the Nazi euthanasia program, Michael Burleigh's *Death and Deliverance: "Euthanasia" in Germany 1900–1994*, was published in 1994 almost 50 years after the end of World War II. This is parallel to Watson's piece shedding light on the euthanasia of the King of England, 50 years after the fact. While writing as to the former was perhaps too horrifying to catalogue in full sooner, and writing as to the latter was perhaps too discreet, the 50-year lag is of interest. The "why's" may be different, but the "when's" were surprisingly similar (although there had been many a discussion of the Nazi atrocities over the years).

It remains interesting that "it took Professor Arthur Caplan, now Director of the Center for Bioethics at the University of Pennsylvania Medical Center, more than a decade to organize a 2-day conference to examine 'The Meaning of the Holocaust for Bioethics' at the University of Minnesota, Center for Biomedical Ethics."[41] For Caplan, the question was why "bioethics had paid so little attention to the obvious dilemma raised by the reality that Nazi doctors and scientists had grounded their actions [generally, and certainly with regard to euthanasia] in moral language and ethical justifications."[42] Both Caplan and Burleigh were seeking to fill what Caplan called a "huge and inexcusable gap in the literature."[43]

Some contrasts between the Anglo-American movement and the German program are bright line. First, Aktion T-4 was a program for "children's euthanasia." While Burleigh notes that "the exact origins of the 'euthanasia' programme are complex, there being several versions of how it started,"[44] as an intellectual matter, he attributes the spark to the winter of 1938–1939, when "the parents of a malformed child called Knauer petitioned Hitler in order to bring about its death."[45] Burleigh's contention was that "this was probably not an isolated incident . . . [i]n such a political climate, understandable human anxieties about severely disabled infants, terminal illness and severe incapacity were compounded by grass-roots ideological fanaticism."[46] As an initial matter, this demonstrates a cultural willingness for families to abjure their familial responsibilities, as well as to use the very suffering St. John-Stevas was critical of, in order to allow for non-voluntary euthanasia (by definition, since an infant does not have capacity to consent, even if of a normal or superior intellectual ability).

In any event, the T-4 program allowed for the possibility of euthanasia in cases precipitating the movement for (and deriving the benefit of) euthanasia as a "blessing of [which] was only to be granted to [true] Germans."[47] What developed into the villainous "euthanasia Aktion T4" program for the killing of the mentally ill and handicapped in 1940 and 1941, and now-infamous "Aktion f 3," "which led to the involuntary 'medical' euthanasia of some 275,000 people, were actually—as a matter of law—criminal in nature, whereas voluntary euthanasia, as originally contemplated, had been rejected by the Nazi dominated Reichstag in 1933."[48]

Although in violation of the national penal code, it became a nationwide policy to administer euthanasia to those deemed to be mentally defective, psychotics, epileptics, and those "suffering" from old age and its related infirmities, as well as organic neurological disorders. This was later extended to Jews, foreigners, members of other races, homosexuals, and other "impure" people. The minister of justice (unsuccessfully) demanded cessation. However, "the German doctors and medical students who opposed the Nationalist Socialist party, those who were unable to flee into the army to remove themselves from the conflict of duties found themselves to be facing 'elimination.' "[49]

Many doctors were alleged to have participated, with impunity (not indicted at Nuremberg), and one review of cases determined that doctors in fact "profited from the unique opportunity to experiment on living human beings, and they supported the Nazi utopian view of a society cleansed of everything sick, alien, and disturbing."[50] This draws attention to what might happen notwithstanding that the doctors were described as " 'average' physicians in terms of their attitude, thinking and daily routine, whose diaries and journals came to light only in the late 1970s and early 1980s (again reflecting nearly a half-century of tacit moratorium), which stirred great debate among members of the German medical profession."[51] Right-to-die advocate Derek Humphry and coauthor Ann Wickett observed (as a critique) that there was no record of the Nazi doctors either killing or assisting in the suicide of a patient who was suffering intolerably from a fatal illness.[52] While that does not mean that there were no compassionately motivated acts of medical euthanasia, it is stunning that these would have gone unreported, while non-voluntary and involuntary medical euthanasias were reported. One must ask a rhetorical question: why would those who were ill and suffering and potentially requesting relief via medical death be unreported, perhaps unreportable, while non-voluntary and involuntary euthanasia was reported upon in detail? The only seemingly possible answers to such a question point to, indeed underscore, the slippery-slope argument.

There was no question that as a result of the Nazi euthanasia program, of non-voluntary and involuntary euthanasia, a number of changes in social attitudes occurred. In England, the common person became more sympathetic to disabilities and to the disabled generally.[53] Moreover, in Anglo-American culture, the Nazi program was not considered to be euthanasia, as it was a "code name" for removing the politically valueless, deformed, insane, senile, or anyone else viewed as politically unsavory.[54] However, perhaps the greatest reflection of how the Nazi program influenced the English was that from 1944 to 1946, there was very little British press on euthanasia and the movement at all. Across the ocean, in 1946, 1,776 physicians joined a committee in New York State to seek lawful voluntary euthanasia, and the next year, "the majority signed a petition to the legislature to amend the law."[55] The petition was not successful, and euthanasia remained a crime under the homicide provisions, which were tested in 1991 by Timothy Quill (who was prosecuted for assisting in a suicide by providing a prescription for a lethal dose, though he did not administer the drugs himself).

CONCLUSION

An unlikely series of juxtapositions comes of the UK (and U.S.) experiences and responses to medical euthanasia in the 1930s and 1940s to that of the Nazi Germany atrocities. As an initial matter, there is a similarity to address. That it took 50 years after the 1936 death of King George V for Watson to write for an academic mass (and speak to the general press) about Lord Dawson of Penn's euthanasia of the monarch, and that it took the same 50 years (until 1994) for Burleigh to conduct the first full-fledged study of the Nazi euthanasia program, leads to the conclusion that matters of euthanasia were unspeakable—whether single crime or mass genocide. Caplan's well-considered contemplation that it took him a decade to organize a conference (nearly contemporaneously to the Watson and Burleigh writings) about the various dilemmas raised by the Nazi euthanasia program among bioethicists, rather than simply fill a "huge and inexcusable gap in the literature,"[56] may be met by an alternative explanation.

That parallel possibility is to say that perhaps the gaps—both as to the taking of the English monarch's life by his personal physician (and the underground practices that doctors admitted to in the 1930s), and as to the taking of the European lives lost to the Germans, passed beyond the medical boundary of *primum no nocero* (first do no harm) into a sociological one, regarding "death workers' " general reluctance to engage in conversation about their work. This sort of finding has been experienced by others in far less controversial (and far more socially, publicly supportive) matters such as funeral and mortuary work, as discussed by British professor Glennys Howarth, who had difficulty getting funeral directors in London to speak, even in confidence, about their work, as she conducted field work for her 1991 doctoral thesis.[57] Indeed, William E. Thompson, writing contemporaneously with Howarth, made similar observations in "Handling the Stigma of Handling the Dead: Morticians and Funeral Directors."[58] The taboos and stigmas associated with death work generally are amplified in an illegal practice (euthanasia in England) and a warping of the law (the Nazi euthanasia program). As an aside, these amplified (indeed, geometrically multiplied) taboos and stigmas were perhaps best shown by the silences of the euthanasia debate in the United States and Britain during the 1940s (after World War II and the Nazi atrocities started and ended). As a second aside, but also expressing the in-the-shadows experience as to death workers (generally) and euthanasia/assisted suicide (specifically) is that half a century later, in 1990s Michigan, among Jack "Dr. Death" Kevorkian's greatest adversarial foes was Senator Fred Dillingham, who had been a mortician prior to his legislative career, a high point of which was his role in criminalizing assisted suicide.[59] These comments about "silences" add perspective, serve to augment and to supplement, or perhaps to underscore, juxtaposed matters that were public and debated during the period in question.

First, in England, discussion and debate became more prominent after the illegal (but not discussed) euthanasia of the King, in which the Queen and royal family were complicit. In Germany, discussion and debate were held to precede legislation permitting euthanasia, and what became a slippery slope toward genocide and war crimes.

Second, the euthanasia debate in England was quelled less by morality matters, than by medical matters—that the doctor should be permitted to proceed, unimpeded, in his (as almost all doctors then were men) decisions as to what treatment was in the best interest of the patient, according to the best analysis by the doctor. After all, doctors who had in the past been left to watch helplessly to watch (or, as in the case of George Washington, to inadvertently hasten) the deaths of their patients, benefitted from (and was emboldened by) the modern medicine, with its treatments and cures, and the more interactive relationship between the medical profession and its clientele. In contrast, while the Nazi euthanasia program may have been to relieve parents of the burdens of defective, mentally or physically impaired infants, and extended to mentally and morally defective adults (as determined by the Nazis), the focus seemed to be what was best for German society (rather than the patient), what had been a thin edge of a legal wedge degenerated to a slippery slope of systematic abuse.

Third, illegal practices (ironically as to both medical practice and unlawful practice) by doctors in both the United States and the United Kingdom became a private matter—ungoverned by any system, medical or legal. While it took nearly 20 years (as will be discussed in the next chapter) for doctors in England and the United States to face public discipline by the legal and medical systems for their acts in hastening deaths of patients, perhaps the most surprising fact was that the doctors were caught out in systems by others who were essentially in subordinated positions to the doctors. Implicitly (and apparently tacitly), doctors in the wake of the Anglo-American debates and the Nazi experience stayed quietly out of the public eye, and conducted themselves within the then-developed (but very private) sphere of the doctor/patient/family, rather than that of public debate or medico/legal attention seeking behavior.

NOTES

1. Hoefler, James M., and Kamoie, Brian E., *Deathright: Culture, Medicine, Politics and the Right to Die* (Boulder, CO: Westview Press, 1994), 46–47.
2. Barrington, Mary Rose, "Euthanasia: An English Perspective," in Arthur Berger and Joyce Berger, eds., *To Die or Not to Die: Cross-Disciplinary, Cultural and Legal Perspectives on the Right to Choose Death* (New York: Praeger, 1990), 97.
3. Lelyveld, Joseph, "1936 Secret Is Out: Doctor Sped George V's Death," *New York Times*, November 27, 1986, http://www.nyt.com (accessed November 28, 2011).
4. Pappas, Demetra M., "Recent Historical Perspectives Regarding Medical Euthanasia and Physician Assisted Suicide," *British Medical Bulletin* 52, no. 2 (1996): 386–93; p. 387, footnote 4, and accompanying text, *citing* Hoefler and Kamoie, *op. cit.*, 46–47.
5. *Id.*
6. Lelyveld, *op. cit.*
7. Watson, Francis, "The Death of George V," *History Today* 36, no. 12 (December 1986): 21–31.
8. Lelyveld, *op. cit.*
9. Watson, *op. cit.*, 28.
10. *Id.*
11. *Id.*
12. *Id.*

13. *Id.*
14. *Id.*
15. Kamisar, Yale, "Some Non-Religious Views against Proposed 'Mercy Killing' Legislation," *Minnesota Law Review* 42, no. 6 (May 1958): 969–1042; p. 971.
16. *Id.*, citation omitted.
17. Williams, Glanville, *The Sanctity of Life and the Criminal Law* (New York: Alfred A. Knopf, 1957), 328.
18. N. D. A. Kemp, *"Merciful Release": The History of the British Euthanasia Movement* (Manchester: Manchester University Press, 2002), 88.
19. Kamisar, *op. cit.*, 1036.
20. Kemp, *op. cit.*, 88.
21. Williams, *op. cit.*, 334–35.
22. *Id.*
23. Kamisar, *op. cit.*, 978, reciting Section 2 (1) of the English bill.
24. *Id.*, referring readers to the full text of Roberts, *Euthanasia and Other Aspects of Life and Death* (1936), 21–26.
25. Watson, *op. cit.*, 28.
26. *Id.*
27. Williams, *op. cit.*, 338, quoting.
28. Williams, *op. cit.*, 338, quoting.
29. Kemp, *op. cit.*, 92.
30. Kamisar, *op. cit.*, 981.
31. *Id.*, 984.
32. St. John-Stevas, Norman, *Life, Death, and the Law: Law and Christian Morals in England and the United States* (Bloomington: Indiana University Press, 1961), 273.
33. St. John-Stevas, *op. cit.*, 266.
34. Hilliard, Bryan, "The Moral and Legal Status of Physician-Assisted Death: Quality of Life and the Patient Physician Relationship," *Issues in Integrative Studies* (2000).
35. St. John-Stevas, *op. cit.*, 277.
36. Kamisar, *op. cit.*, 988, quoting 103 H.L. Deb. (5th ser.) 466, 492–93 (1936).
37. Kamisar, *op. cit.*, 995, quoting Wolbarst (1935).
38. St. John-Stevas, *op. cit.*, 273.
39. Kamisar, Yale, "Euthanasia Legislation: Some Non-Religious Objections," in A. B. Downing and B. Smoker, eds., *Voluntary Euthanasia: Experts Debate the Right to Die* (London: Peter Owen, 1986), 110.
40. Lamb, David, *Down the Slippery Slope: Arguments in Applied Ethics* (London: Croon Helm, 1998), 11.
41. Pappas, *op. cit.*, 390–91, *citing* Arthur L. Caplan, "Preface," in *When Medicine Went Mad: Bioethics and the Holocaust* (Totowa, NJ: Humana Press, 1992), vi.
42. *Id., citing* Caplan, *op. cit.*
43. *Id.*

44. Burleigh, Michael, *Death and Deliverance: 'Euthanasia' in Germany 1900–1945* (Cambridge: Cambridge University Press, 1994), 93–94.

45. *Id.*, 93.

46. *Id.*

47. Kamisar, *op. cit.*, 140.

48. Pappas, *op. cit.*, 390.

49. *Id.*, *citing* British Medical Association, *Euthanasia* (1988; *citing* H. M. Hanauske-Abel, "From Nazi Holocaust to Nuclear Holocaust: A Lesson to Learn?" *Lancet* [1986], ii, pp. 271–73).

50. Pross, C., "Nazi Doctors, German Medicine, and Historical Truth," in G. J. Annas and M. A. Grodin, *The Nazi Doctors and the Nuremberg Code: Human Rights in Human Experimentation.*

51. Pappas, *op. cit.*, 390.

52. Humphry, Derek, and Ann Wickett, *The Right to Die: Understanding Euthanasia* (New York: Harper & Row, 1986), 23.

53. Kemp, *op. cit.*, p. 119.

54. *Id.*, p. 44.

55. St. John-Stevas, *op. cit.*, 266.

56. Pappas, *op. cit.*

57. Indeed, this secrecy, and the difficulty in getting discussion of this relatively benign practice, was the subject of a chapter of Howarth's 1991 dissertation for the London School of Economics, subsequently published as a book entitled *Last Rites: The Work of the Modern Funeral Director* (New York: Oxford University Press), 38, and generally, 32–52.

58. Thompson, William E., "Handling the Stigma of Handling the Dead: Morticians and Funeral Directors," *Deviant Behavior: An Interdisciplinary Journal*, no. 12 (1991): 403–29.

59. Pappas, Demetra M., interview with Senator Fred Dillingham (Michigan), August 23, 1993.

3

The 1950s: The First Anglo-American Prosecutions for Medical Euthanasia and the Resulting Academic Debate

During the 1950s, medical euthanasia went from being an abstract debate about an in-the-shadows practice of doctors to an open matter of public record. That is because the first two defining prosecutions of doctors took place, although neither was charged and tried for euthanasia per se. In addition, the legal system saw the first cases and legal standards regarding related matters such as what is (or is not) informed consent, and what the definition of death was. These came together to prompt three of the greatest minds of an Anglo-American legal generation to present written arguments regarding the interplay of the sanctity of life, the principle of autonomy for the terminally ill, and both religious and secular objections to euthanasia.

By the end of the decade, the groundwork was laid for the seeming legal fictional defense that causation (a key element of any murder charge and trial) could not be proven, because the decedent (or victim) could have died from an underlying illness; and for a second legal defense (either fictive or not) that the intent of the defendant (also a key element of any homicide charge and trial) could have been benign, rather than criminal in nature. Either of these possibilities, if successfully raised, creates a reasonable doubt that the crime of homicide has been committed, resulting in acquittal on such a charge.

THE 1950 TRIAL OF DR. HERMANN SANDER: CREATING THE CAUSATION ISSUE OF "ALREADY DEAD"

The decade literally opened with the New Hampshire case against Dr. Hermann Sander, who was charged with first-degree murder relating to the December 4, 1949, death of a 59-year-old cancer patient.[1] These charges of injecting the decedent with four air bubbles, in close succession, were treated with great seriousness, requiring a $25,000 bond (raised by friends) pending trial, and a possible sentence of capital punishment (death by hanging) or life in prison, if Sander was convicted.[2] The grand jury specifically found that Sander administered the injections "well knowing" that these were "sufficient to cause death."[3] The judge presiding over the grand jury warned that "the law 'must be enforced as it stands,' and that 'it is not up to you to say what the law should be.' "[4] Even as Sander was being indicted, the Euthanasia Society of America announced that it would seek the first "mercy-death law" in the United States; this would be by an amendment to a law allowing for voluntary euthanasia (which it described as legalized mercy killing) for incurable patients and "sufferers upon their petition and with the recommendation of a medical committee and the approval of the courts."[5]

Dr. Sander remained at liberty pending the outcome of his case, because of "his high moral character and standing in the community."[6] In addition, Sander agreed to suspend his practice during the pendency of the case.[7] Media and public support rallied around Sander during this period, with an anonymous piece in the *New York Times* criticizing a legal system in which mercy killing was deemed to be the same as murder, and in which defendants resorted to juries, which refused to convict for euthanasia.[8] Sander continued to enjoy the respect of the medical community, as a director of the New England chapter of the American Academy of General Practice, even as he went on trial for murder.[9] Although the Academy considered seeking the expulsion of Dr. Sander and introducing a resolution opposing euthanasia and Dr. Sander, on February 20, 1950, Mac F. Cahal, the executive secretary of the Academy's Wisconsin chapter, said:

> There are extremists on both sides . . . [who] will bring to the floor emotions more than anything else. The subject is emotional at present. It is a moral question, not a scientific one. It is a matter for the church, not for doctors.[10]

With this statement, the discussion went from one regarding murder to a debate of morals and theology versus medical mercy, and it also predicted the sort of extremism and zealotry that would continue to develop, and that ultimately would mark the 1990s trials of Jack "Dr. Death" Kevorkian. However, Sander himself was a somewhat reluctant protagonist who reputedly "did not deliberately go to work to bring about a test case. He said it never occurred to him that it would cause trouble when he honestly made a notation on the hospital records of what he had done."[11] However, other doctors, such as Dr. A. L. Goldwater of New York, did, in fact, go on the record at meetings of the Euthanasia Society of America, to state that they had made overdoses of morphine available to "incurable patients."[12] Goldwater described in detail that it was "his custom in an incurable case was to prescribe a week's supply of morphine and leave them on the patient's bedside table," with instructions.[13] Goldwater further claimed that this was a practice common to many physicians. This practice, described nearly half a century before anyone heard of Dr. Jack Kevorkian or Dr. Timothy Quill, was, for all intents and purposes, what in the twenty-first century became lawful assisted suicide in Oregon and Washington.

At trial, the defense planned to introduce evidence that the 41-year-old doctor was depicted by witnesses as a kind, compassionate doctor, who watched with horror as Mrs. Borroto wasted away from cancer for one and one-half years, going from 140 pounds to a mere 80, and spent her final two weeks in the hospital.[14] In the days before the trial, information also was published regarding the prosecution's case. Specifically, the root of the prosecution seemed to be notes that Sander had dictated to a nurse (who was also the hospital records librarian) indicating that he had injected Mrs. Borroto with air bubbles, which the nurse reported to the hospital chief of staff, who in turn made a report to the police.[15] Sander said that administering the air bubbles was "an act of mercy, without malice" out of "pity for his patient and desire to end her suffering . . . she would probably have died of cancer within a few hours."[16] County officials ruled the cause of death to be cancer, as the amount of the air bubble was, they contended, too small to cause death.[17] In addition, the prosecutor, William Phinney, who had gone to school with Sander, pointedly commended Sander's "high moral character," despite the prosecution.[18] This almost anticipated similar phrases and comments by the Wayne County chief prosecutor in 1993, when indicting Dr. Jack Kevorkian in the first assisted-suicide case that would go to trial in Michigan; likewise, the presence of some 50 reporters seemed to predict the reaction of the press to the Kevorkian cases in the 1990s.[19]

Anticipating the press, the trial judge withheld the names of potential veniremen (jurors) and planned a pool of 160 potential jurors (four times the ordinary size of the jury pool), so as to ensure an impartial jury, a sentiment that would be echoed in the 1990s during selection of the Kevorkian juries.[20] Moreover, the court planned to secure the jury in a local hotel when not in court hearing (or deliberating) on evidence.[21]

During the trial, some 23 men and women, including "old and young, rich and poor," came forth to testify as to their admiration of Sander, and doctors, nurses, patients and friends" came forth to defend him.[22] One such witness, Dr. Robert Rix, testified that "Dr. Sander at the time of the killing [*sic*] had shown signs of mental and physical fatigue and had seemed 'a little' high strung and over-wrought. [Dr. Rix] said his own wife had died of cancer and Dr. Sander was one of the pallbearers on Dec. 3, the day before he injected the air."[23] All the doctors "were forced to admit that air injections in a vein have no therapeutic value, do note alleviate pain, and are not part of any medical treatment, but on the contrary are known to cause death if given in sufficient quantities."[24]

Sander testified in his own defense at trial, acknowledging that he administered four injections of air bubbles into Mrs. Borroto's veins, about which he testified in detail, but emphatically stating that he "never had any intention to kill" her.[25] The essence of the detailed medical testimony was that there was no backflow of blood from Mrs. Borroto's veins, which should happen when an injection is administered to a living person. In addition to hearing all of this, the 12-man jury (plus one alternate juror) visited the scene of Mrs. Borroto's death and had a tour of the hospital where it had occurred.[26]

In the midst of all of this, local Reverend C. Leslie Curtice, a pastor and "close friend" of Dr. Sander, "set the tone for the psychological aspect of... his defense in the 'mercy killing' trial."[27] Specifically, the pastor was quoted as saying, "when a man devotes himself conscientiously to his task and fulfills the spirit that goes beyond mere observance of the law to faithfully carry out deeds of kindness and good will, as Dr. Sander did, then you have a genuine man of virtue, a religious person."[28] This was in the course of a sermon in which he encouraged the donation of funds to Sander's defense fund.[29] Other expressions of community faith in the doctor included a petition signed by 605 of his town's 650 residents, as a "statement of their faith in him."[30]

While this community defense of Sander was taking place, Sander's lawyers effectively argued that his case was not one of euthanasia or "justified" mercy killing, but rather that no murder had been committed because the "victim" had died of natural causes.[31] Thus this case that began as an opportunity as a test case to debate euthanasia ended as a simple case tried on the facts, and which resulted in an acquittal after only 70 minutes.[32]

The New Hampshire State Board of Registration temporarily revoked Sander's license, subsequent to his acquittal (but reinstated it, barely two months later), because the trial raised the issue of medical euthanasia.[33] This decision suggests that it was not the actual conduct that the New Hampshire State Board of Registration found objectionable, but rather the trial and attendant publicity (this sort of thinking would accompany the final trial of Jack Kevorkian, in 1999). However, the New Hampshire State Medical Society (the state professional association) held a three-hour closed-door session, the result of which was a decision not to discipline him.[34] This was despite his admission on the witness stand that he had injected air into her arm, but claimed that she was already dead by then. Sander's explanation "that the injection was given in an 'irrational' moment when 'something snapped' in him after he had witnessed for many months her great pain and suffering,"[35] may have convinced the jury, but astonishingly was not viewed as a violation of his oath as a doctor.

In specific, the New Hampshire State Medical Society stated that, "the development of the case and the verdict of the jury have established that mercy killing was not involved. A contrary impression was created by the publicity associated with the case."[36] Sander, who had not practiced since his indictment on January 3, had his license temporarily suspended, from April 19, 1950 until June 28, 1950; upon his reinstatement. Sander was receiving general practice patients the same day.[37] Noted was that "international interest centered on Dr. Sander's trial, which started Feb. 20 and ended March 9, because it was expected to be a test case on euthanasia, mercy killing. That question, however, never became a trial issue."[38]

Ultimately, the prosecution and trial of Dr. Hermann Sander provided an opening for medical staff (primarily physicians) who might be accused of euthanasia. What was essentially a legal fiction—that the patient died of the underlying illness between the time of a lethal injection of either drugs or air bubbles, and the time the injection would have stopped the heart—is one that a jury could hang a hat on in concluding that there was a reasonable doubt as to the cause of death. Indeed, while this seems to be incredible, this was exactly what a jury found in Winchester Crown Court in 1992, in the prosecution of Dr. Nigel Cox, who was accused of injecting two lethal ampoules of potassium chloride (which stops the heart) into the veins of an elderly patient who was terminally ill with rheumatoid arthritis. The patient, 70-year-old Lillian Boyes, was in the hospital at the time, and her bones could be heard audibly moving when people touched her. After her death, her son Patrick thanked Cox for his assistance, which was believed to have shortened the dying woman's life by hours or perhaps days. Cox wrote notes in the file to the effect that he had administered the lethal drug, which a nurse saw, and reported.

As an irony, the 1992 trial was successfully defended on the ground that the patient was already dead; a throwback to the 1950s. Dr. Cox, while not convicted of murder, was convicted of attempted murder, and remains the only physician to have been convicted of attempted euthanasia in the United Kingdom. A central reason for the 1992 attempted murder conviction is that the body of the decedent had been cremated before an autopsy could be performed. Thus, cause of death could not be ascertained. In turn, only an attempt could be proven beyond a reasonable doubt—with the crucial damning evidence of Cox's notes in the hospital record.

This all said, Cox did not serve a single day of his one-year suspended sentence after his trial. The disciplinary comments during sentencing were more in keeping with a petty theft than a major felony. Indeed, the General

Medical Council, rather than banning Cox from practice, required him to take courses in palliative care

The Cox case sparked considerable debate, and was one of the two prompts for the 1993–1994 House of Lords Select Committee on Medical Ethics. However, insofar as the Cox case was concerned, the most frequently heard comment in the Select Committee hearings was surprise—not that Cox had engaged in the conduct, but that he had written it down, and left an evidentiary trail for hospital administrators and the Crown Prosecution Service.[39] Although the Cox case and the House of Lords Select Committee on Medical Ethics were deliberated upon in the 1990s, they had as a root the 1950 case against Hermann Sander.

What is critical to note is that the trial of Dr. Hermann Sander was not on the issue of euthanasia, but rather was on the elemental question of whether the decedent, Mrs. Borroto, was already dead at the time he injected her with the air bubbles. Indeed, defense attorney Louis B. Wyman emphatically stated—after the acquittal—"euthanasia is not the defense. . . . [w]e haven't raised it. We say the doctor is not guilty of any malicious killing."[40] Once this question was decided, and the jury determined that she was already dead, the jury did not go on to deliberate as to whether Sander attempted to commit murder. In later cases, such as the Cox case, the attempt charge was deemed a lesser included offense of murder. While the sentencing judge chose not to sentence Cox to prison, and rather imposed a single-year suspended sentence, the risk of criminal (and civil) liability continues to loom large in cases in which the defense is that the patient is "already dead."

THE 1957 TRIAL OF DR. JOHN BODKIN ADAMS: CREATING THE "DOUBLE EFFECT" DEFENSE

The next definitive case of the 1950s, that of England's Dr. John Bodkin Adams, produced a defense leading to full exoneration is that of "double effect," or that the doctor intended to alleviate pain and suffering, with an "unintended" consequence of shortening life. Several decades after the trial, the trial judge, Lord Patrick Devlin, published a book, *Easing the Passing: The Trial of Dr. John Bodkin Adams,*[41] which was for all intents and purposes, a trial memoir. Such a memoir was, itself, highly unusual at the time (and remains so). Lord Devlin, in acknowledging this, correctly observed that:

> the lapse of a quarter of a century is "long enough to allow for publication without indecorum. . . ." This view has been generally, but not universally, accepted. . . . Certainly he must not write about his cases while he is on the bench.[42]

In this first murder trial of a doctor for allegedly killing a patient, Adams was charged with the November 13, 1950, murder of Edith Alice Morrell,[43] to which he pleaded not guilty.[44] The anesthetist prescribed a maximum dosage of morphine and 75 percent above the maximum dosage of heroin for his patient, who was suffering from the effects of a stroke, but allegedly not suffering from pain.[45] One reason that the case came to the attention of prosecutors was that on July 18, 1950, Mrs. Morrell executed codicils to her will in July and August 1950, of which Dr. Adams was the beneficiary.[46] This hints at an issue that would become prominent when assisted suicide became lawful in the 1990s in Oregon and later in Washington State—the procedural safeguard that no physician involved with an assisted suicide of a patient could be a beneficiary of the patient's will.

Unlike Sander, Bodkin Adams did not administer the injections in question, but rather wrote orders for a nurse to do so.[47] While the nurse "did not like" giving repeated and very large injections, and "Dr. Adams having given her those instructions—of course it was her duty to do so."[48] During a subsequent inquiry by New Scotland Yard, less than two weeks later, Bodkin Adams told the detective superintendent:

> Easing the passing of a dying person is not all that wicked. She wanted to die. That cannot be murder. It is impossible to accuse a doctor.[49]

This statement and argument, along with the statements of Dr. Bodkin Adams on December 19, 1950, upon his arrest at his home, almost predicted some of what Jack "Dr. Death" Kevorkian and his lawyers would argue during the 1990s:

> **Dr. Bodkin Adams:** Can you prove it was murder?
> **Supt.:** You are now charged with murdering her.
> **Dr.:** I did not think you could prove murder. She was dying in any event.[50]

Thus, whether Bodkin Adams was acting upon a financial or a compassionate motive, he was not disputing the element of causation—that is, what the medical (and legal) causes of Mrs. Morrell's death would be found to be. As the first doctor to be tried in Britain for a murder relating to euthanasia, and unlike Sander, Bodkin Adams did not take the stand at trial. That is not to say that administering opiates was without medical discussion. Indeed, it had been long documented that morphine could shorten life, even if administered for the purpose of relieving pain. For instance, consider this 1887 argument that:

> Opium is administered to the dying . . . should rarely be administered to the dying as mere hypnotic or with a view to enforce sleep. To do so would be to risk throwing the patient into a sleep from which he may not wake.[51]

Cause of death having been removed from the issue meant that the actus reas, or the guilty act, was removed from issue. Thus, the central question at trial was whether Bodkin Adams possessed the mens rea, the guilty mind or the requisite intent, to kill his patient. Instructing the jury, Lord Devlin charged the jury that if the treatment was medically appropriate (for instance, to relieve pain), but that death was hastened as a result of the medication, this secondary effect was not within the ambit of a murder charge.[52] In specific, Lord Devlin instructed that while "murder is the cutting short of life, whether by years, months of weeks [and that] it does not matter that Mrs. Morrell's days were numbered,"[53] a doctor was not required to "calculate in minutes or even hours, and perhaps not in days or weeks, the effect upon a patient's life of the medicines which he administers or else be in peril of murder."[54] This opened the door to what was next to follow, which became the standard for the defense of "double effect":

> If the first purpose of medicine, the restoration of health, can no longer be achieved, there is still much for a doctor to do, and he is entitled to do all that is proper and necessary to relive pain and suffering, even if the measures he takes may incidentally shorten life.[55]

Devlin went on to elucidate for the jury that:

> This is not because there is a special defence for medical men but [*sic*] because no act is murder which does not cause death. We are not dealing here with philosophical or technical cause, but with the commonsense [*sic*] cause.[56]

In point of fact (and of law), the "technical" cause is actually quite important in determining cause of death. While the jurors are indeed supposed to bring their common sense and good judgment into the jury room, the judge is supposed to instruct the jury as to the elements of the law to apply to the facts of the case. These elements include causation of the decedent's death and intention of and relating to the defendant's act. Thus, Lord Devlin's instruction to the jury was that "the cause of death is the illness or the injury, and the proper medical treatment that is administered and that has an incidental effect on determining the exact moment of death is not the cause of death in any sensible use of the term."[57] While Devlin also intoned that "no doctor, nor any man, no more in the case of the dying, than of the healthy, has the right deliberately to cut the thread of life,"[58] the preceding and specific instructions as to what came to be known as "double effect" resulted in a verdict of not guilty on April 9, 1957, in a spare 44 minutes.[59] In short, the Bodkin Adams jury took even less time to deliberate than the Sander jury to acquit, with both approximating one mere hour of deliberations. When one considers that the jury retired to deliberate at 11:16 a.m.,[60] and returned with a not guilty verdict at 12:00 noon,[61] one may reasonably wonder if the jurors had a civilized cup of tea, consistent with the British custom of "elevenses," the equivalent to an American morning coffee break, and then returned with their verdict; an American counterpart to this has been observed by criminal lawyers where a jury retires to deliberate before lunch, and returns a verdict immediately after the luncheon break (with a verdict of either guilty or not guilty). There can be no doubt that the attorney general, Sir Reginald Manningham-Buller Q.C.,[62] thought so, with regard to a pending case against Bodkin Adams regarding the death of one Mrs. Hullett; the attorney general "reached the conclusion that in all the circumstances, the public interest does not require that Doctor Adams should undergo the ordeal of a further trial on a charge of murder."[63] The attorney general almost predicted the reaction of David Gorcyca, the Oakland County Michigan chief prosecuting attorney who dismissed numerous assisted-suicide charges against Jack "Dr. Death" Kevorkian in the 1990s, because he thought these would result in acquittal (all prior to the infamous 1998 CBS *60 Minutes* "Death by Doctor" segment showing Kevorkian's euthanasia of Tom Youk, which led to Kevorkian's conviction by Gorcyca's office). A full half-century before Gorcyca took office, at the end of the Bodkin Adams trial, the attorney general instantly (indeed, on the same transcript page) cited the fact that the press attendant to the Bodkin Adams trial would inevitably lead to difficulty obtaining a jury, indicted that he did not seek to repeat the process of a lengthy trial that resulted in acquittal, and expressed dismay over the fact that evidence of admissions did not seem to have been given great support in the case.

Lord Devlin, in an addendum to *Easing the Passing*, conceded that after the Bodkin Adams trial, "and no doubt as a result of the Adams trial," the law was changed so that doctors "who kept dangerous drugs for use in his practice must keep a register showing how they were acquired *and how disposed of*" (emphasis original).[64] Not often mentioned is that Bodkin Adams was also accused of murdering other patients, in whose deaths he had a beneficial interest and in which cases he pleaded guilty to minor charges, and was struck from the medical register. That all said, the principle of double effect became a time-honored defense in assisted-suicide and euthanasia trials and in legislative enactments regarding assisted suicide, as will be developed in further chapters.

A COMPARATIVE STUDY OF SANDER AND BODKIN ADAMS, AND THEIR CASES

A juxtaposition of these two cases reveals differences in the doctor defendants, prosecutorial discretion in how to treat the cases, the defenses, and community responses.

First, Sander was viewed as a young doctor, who became overwhelmed (whether as a legal fiction or as a matter of fact) by deaths around him, as well as by his patient's decline. In contrast, Bodkin Adams was perceived as a doctor who was not in the least overwhelmed, but rather seemed to be motivated by a financial interest. This, in turn, leads to the quite reasonable concern that Bodkin Adams was taking advantage of his vulnerable patient (and indeed, over time, *patients*).

The concerns raised in this latter regard of the Bodkin Adams case would persist throughout the twentieth century and beyond. A central concern for legislatures considering whether (and how) to decriminalize, legalize, and regulate assisted suicide so as to protect vulnerable populations, such as the aged, ethnic minorities, the marginalized poor, and with regard to gender disparities. While Bodkin Adams was apparently involved in hastening the deaths of wealthy people for his own financial benefit, it cannot be overlooked that the original Nazi euthanasia program was to benefit the elite, and only later was bastardized for ethnic cleansing and killing members of marginalized and vulnerable populations.

Second, Sander had the support of the decedent's family before, during, and following the trial. While a motive of mercy does not mitigate the element of intent, and family support of a doctor defendant should not (but often does) lead to the path toward mitigation (if not outright exoneration), the converse is true—a motive of money, of financial gain, may be viewed (in fact, if not in law) to be indicia of intent. Moreover, that there was an impact on the family of the decedent in the Bodkin Adams case increased the likelihood of family agitation in the case. There is a credible argument that if Bodkin Adams had not had a financial interest in the case, and pursued it with aggressive zeal, the death of Mrs. Morrell would have been quietly forgotten, rather than subject to investigation and prosecution. Indeed, Lord Devlin's outline of Bodkin Adams's life and his legal difficulties regarding other such cases of financial interest clearly points to this conclusion.

Third, Sander was tried for one case of murder of one patient, albeit under a theory of extenuating circumstances. While the prosecutor was clear that shortening life of even one patient, even briefly, is murder, Sander was permitted to bring in all manner of evidence of mitigation. Of course, mitigation is not a justification or a defense of murder, but it can be used as an excuse—in sentencing, once a conviction upon a verdict of guilty has already been secured. In contrast, with Bodkin Adams, there was more than a whiff of a persistent pattern of criminal behavior motivated by something akin to larceny by trick, with a deadly outcome; indeed, he ultimately pleaded guilty to lesser charges in another case.

Fourth, the defense in the Sander case never challenged the intent, gambling that what was almost certainly a legal fiction, that the patient was "already dead" when he injected her, provided the jury with a nail to hang the hat of jury nullification (acquittal in the jury's perception of the interests of justice, despite proof of the elements beyond a reasonable doubt). The defense in the Bodkin Adams case had no such option available to deploy as a defense to the crimes charged, given that he ordered a nurse to administer the lethal doses (which she challenged, then reported). The nurse herself was viewed as above reproach for following the doctor's orders, a defense that in the twenty-first century might be less availing due to the more active role of nurses and nurse practitioners (as shall be discussed in Chapter 6, regarding the role of a nurse in the Cox case). Generally overlooked is that a doctor who did something under (arguably) extenuating circumstances is quite different from a doctor who planned (as evidence by the prescription order and nursing instructions) and seemingly used another person (the nurse) to effect his goal.

Fifth, while community response certainly cannot guarantee either acquittal or conviction, the community rallied around Sander, in a way in which Bodkin Adams did not enjoy in his community. If criminal prosecutions are a reflection of the violation of norms in society, then certainly having colleagues defend, churches raise funds, and the family of the alleged victim be grateful for the doctor's treatment, are all indicators of how a jury will perceive a case. Indeed, the parade the townspeople gave Sander following the acquittal is as indicative as anything regarding the community perception of him as an "innocent man," not merely a "not guilty" one. Bodkin Adams had none of this, and had the decedent's family (who lost certain beneficial economic interest to him) challenging his actions and further instigating. While both ultimately faced disciplinary measures by their licensing bodies, another indication of norms (specifically of their professional administrators), they had vastly different outcomes. Sander was unable to practice for a few spare months (with a hue and cry from local townspeople, who called him to consult within 10 minutes of being reinstated). In contrast, Bodkin Adams resigned his licensure from the National Health Service permanently and under pressure, with some disgrace.

There was, however, one area of overlap in both the Sander and Bodkin Adams cases—the speed within which the juries acquitted. That Sander was acquitted in less than an hour and a half seemed a reflection of the community response to him, and to the sympathetic facts of the case. However, the Bodkin Adams verdict, which was returned in a spare 44 minutes, was unquestionably the result of Lord Devlin's instruction regarding primary and secondary effect, what would come to be known as the double effect.

There is one final matter of comparison at this juncture. An acquittal has no precedential effect, by definition. That does not, however, mean that it has no future impact. Rarely has any doctor been tried for a case and asserted the defense that the patient was already dead at the time of an injection or other lethal treatment. (Cox was a rare exception, and the final Kevorkian case, discussed in Chapter 5, offered facts that would have supported such an argument, but that Kevorkian did not argue on summation when he represented himself.)

The defense of double effect, has, however, become part and parcel not only of (largely successful) doctor-defendant defenses, but has become part of the legislative fabric of exoneration and excuse from prosecution (and also a defense to civil liability for malpractice). Thus, the Bodkin Adams case had a permanent impact, despite the unsympathetic facts and even less sympathetic defendant.

HOW IMPRACTICAL REALITIES OF THE 1950s DOCTOR PROSECUTIONS PRECIPITATED LEGAL THEORIES

During the second half of the 1950s, the debate about how the legal system should treat doctors who administer euthanasia to their patients found its way out of the halls of justice and into the halls of academia. There was, in specific, debate among three esteemed legal academics, spread among the United States and the United Kingdom, and derived from both theological and secular points of view. Glanville Williams, writing *The Sanctity of Life and the Criminal Law*, was the first of the three, publishing his arguments against religious (and, in his written view, primarily Roman Catholic) opposition to euthanasia in 1957.[65] Yale Kamisar followed the next year with a seminal article, entitled "Some Non-Religious Views against Proposed 'Mercy-Killing' Legislation."[66] Starting with the very first paragraph, Kamisar took square aim at Williams's arguments, and to reply throughout the article. Last, Norman St. John-Stevas, a Roman Catholic legal philosopher, penned *Life, Death and the Law: Law and Christian Morals in England and the United States*,[67] published in 1961 but presented and printed during the 1950s, too.

Before addressing the various arguments of these three legal legends, it is important to make note of one point relating to the actual trials of Hermann Sander and John Bodkin Adams, and a rare point of unity among the legally arguing trio of the late 1990s. That is, while they refer to the Sander case as giving them reason to consider further, none of them discuss the Bodkin Adams case. This is intriguing because this suggests that they at least agreed on which case presented to them an issue of debate. Moreover, the Sander case was not defended on grounds of euthanasia as either a justification or a defense, or an excuse in mitigation of sentence. Rather, the defense avoided the thorny issues and simply argued a case of legal impossibility—that a dead person, once dead, cannot be killed again or murdered. Yet this was a case that all three legal theorists deemed to further the debate.

The Bodkin Adams case was, however, ignored by all three of the theorists. Placed in historical context, this silence suggests one of two things. Either there was a tacit (or even explicit) agreement that he was beneath their dignity, as one who by June 1957 had resigned from the medical profession in disgrace and pleaded guilty to lesser charges in July 1957 (thereby reducing him to a common criminal). Alternatively, and less likely, they all accepted that the case of Bodkin Adams was not worthy of discussion, notwithstanding the fact that this defense was charged (and successfully so) to a jury for the very first time in the Bodkin Adams case in April 1957 (they did, however, all consider the principle of double effect). This seems to be a paradox, and the silence among the three legal theorists is deafening.

St. John-Stevas noted the then state of the law as to both assisted suicide and euthanasia:

> If an ill person takes his own life, he is guilty of suicide. Doctors or others who assist such persons to take their lives are held responsible as aiders and abettors or principals in the second degree, to the self-murder of another. They are principals if they administer the fatal dose themselves, whether or not the patient has given his consent.[68]

However, all three tacitly accepted as axiomatic what Kamisar stated openly, that whereas "the law on the books condemns all mercy killings,"[69] "[t]he Law in Action is as malleable as The Law on The Books is as uncompromising."[70] Kamisar succinctly described these as failure to indict, acquittal, suspended sentence, and the final (and less often employed) remedy of reprieve.

Perhaps one reason the trio did not discuss the actual trial of Bodkin Adams is that they took the principle of double effect as a given, whether as a practical matter or one of academic debate. For example, Williams, criticizing the Roman Catholic approach requiring the protection of the sanctity of life at all costs, allowed for:

> the medical use of narcotics to annihilate the senses, provided that life is not destroyed. . . . is not euthanasia to give a dying person sedatives merely for the alleviation of pain, even to the extent of depriving the patient of . . . sense and reason.[71]

In a point Kamisar intended to quarrel with Williams (as to the contention that painkillers and palliative care could not relieve all symptoms),

Kamisar argued that, in any event, "the mental side effects of narcotics, unfortunately for anyone wishing to suspend them temporarily without unduly tormenting the patient, appear to outlast the analgesic effects . . . by many hours."[72] What goes unstated here is that a terminally ill patient might not want to have analgesics suspended, and there is not necessarily any reason for a doctor to force a patient into what would later come to be known as withdrawal or cold turkey. St. John-Stevas's contention that there are good drugs was, perhaps, mitigated by his claim that there was an exaggerated claim of patient pain, a subjective measure that he (as a legal theorist) was not in a position to assess. One of his sources was Lord Dawson of Penn, referring to cancer patients.[73] While Lord Dawson may have been a credible source to discuss the level of cancer pain of patients, St. John-Stevas was incomplete as to his description of the physician, who had been King Edward V's personal physician, and had (at the request of the Queen), participated in taking whatever measures necessary to shorten his patient's life in 1936.[74] This suggests that while the King may fully feel pain, and honestly express it, and be medically released from it (i.e., receive euthanasia from his doctor), the common man might perhaps exaggerate it to get more narcotics. This all said, and Williams's suggestion that euthanasia be legalized and regulated in such cases aside,[75] the implicit fact was that any doctors who overprescribe were not prosecuted (Bodkin Adams, tacitly disregarded by all, aside).

This brings up the second category Kamisar referred to—the acquittal possibility. As Kamisar noted, Hermann Sander was the only doctor prosecuted for euthanasia, although another famous American case, *People v. Paight*, which involved a family member engaging in euthanasia, also resulted in acquittal.[76] These cases move beyond those within the contemplation of Roman Catholics, who believe in the sanctity of life, and that life belongs to God alone.[77] This is because a legal acquittal does not negate the intention to medically cause death—with the possible exception of the legal fiction employed in the Sander death, that the patient was already dead. Moreover, Kamisar's third and fourth categories—suspended sentence and reprieve—both require a conviction, placing them outside the ambit of both lawful conduct and Christian ethics.

The heart of the debate of the 1950s legal theorist was, however, whether to allow euthanasia under the law as a permitted course of conduct. This raised a largely secular debate between Williams and Kamisar, with the former arguing in favor and the latter arguing in opposition.

Williams argued for the doctrine of necessity, which "in the common law refers to a choice between competing values, where the ordinary rule has to be departed from in order to avert some greater evil."[78] The greater evil was pain and incurable disease, which would, of course, have Williams leave out a number of the chronic, but not imminently terminal, fatal illnesses that came to the fore in the late twentieth century, such as Huntington's disease, multiple sclerosis, ALS (Lou Gehrig's disease) and end-stage AIDS (which did not even exist as an illness in the 1950s). Williams likened Sander to be in this category (notwithstanding the legally fictive defense asserted).

Williams suggested four possible mechanisms to be charged as murder under the then-existing law; the first was injection. Second, he suggested that a doctor could "furnish poison"[79] to a suicidant. This predicted the development of the assisted-suicide debate. Third, Williams commented that "one may welcome the principle of double effect as an alleviation of the orthodox attitude toward suicide."[80] Williams argued that any of these would have a doctor liable for murder, or "at least manslaughter."[81] Last, Williams noted that mercy killing by omission, conceded "by even the Catholics,"[82] would be presumably exempt from criminal liability (this last all but predicted the right-to-die cases of the 1970s and thereafter). Seemingly as an aside, he observed that "[t]he law is also probably adverse where the physician, in order to put an end to his patient's suffering departs altogether outside any treatment [he] has previously been given."[83]

Williams argued, citing a bit more fully Lord Dawson of Penn, that "there should be no formalities and that everything should be left to the discretion of the doctor . . . by law, and so remove from doctors the fear of the law that now hangs over them."[84] He also argued that "by legalizing euthanasia, one result [is] it would bring the whole subject within ordinary medical practice."[85] Williams also suggested, as if seeing around the bend to the Oregon and Washington legislation to come, that there be requirements for procedural safeguard including a second opinion and patient consent,[86] that such pharmaceuticals be regulated, and that there should be a legislative mandate as to the definition of a doctor.[87]

St. John-Stevas, as if anticipating the 1993 creation of, and 1994 recommendations by, the House of Lords Select Committee on Medical Ethics, suggested that it would be "best to leave the law as is, even if imperfect." This would allow for the malleable application of law as discussed by Kamisar. However, Williams's response was that at the very least, the law should allow for voluntary euthanasia to be a lesser evil, a lesser charge akin to a lesser included offense in murder cases.[88]

However, Kamisar argued against Williams on a number of grounds, which might be put into two categories. First was the possibility of well-meaning error, the possibility of advance toward a cure, and the possibility that psychotherapy might alleviate the wish to die.[89]

Of greater concern to Kamisar was that allowing for euthanasia in any circumstances would be to step onto the thin edge of the wedge, a slippery slope toward abuse. St John-Stevas agreed with this argument, referring to the Nazi experience.[90] Williams, however, derided this argument with one of his own—that soldiers and executioners, who are considered the just agents of a just death, are apparently not subject to this concern.[91]

CONCLUSION

The law itself went unchanged during the 1950s. However, the cases of Dr. Hermann Sander, inviting a causation issue (as well as legal debate), and Dr. John Bodkin Adams (underscoring a medical issue) laid the groundwork for future cases, legislative efforts, and both prosecutions and defenses. What is important to keep in mind is that while Williams was advocating for doctor (and, in turn, patient) autonomy, Kamisar and St. John-Stevas, for different reasons both secular and sanctified, were vigorous opponents for practical as well as theoretical reasons. Whereas Kamisar expressed concern in the erosion of "public confidence in the legal system,"[92] St. John-Stevas pointed to the possibility that allowances for euthanasia would, in addition, "undermine the doctor patient confidence."[93]

There can be no question that the writing of these three theorists all but predicted developments in the 1960s (which will be explored in the next chapter) and the ongoing debate to follow.

NOTES

1. Fenton, John H., "Sander Is Indicted in 'Mercy Killing': New Hampshire Doctor Faces First Degree Charge in a Cancer Patient's Death," *New York Times*, January 4, 1950, p. 12.

2. *Id.*

3. *Id.*

4. *Id.*

5. "Group Will Seek Mercy-Death Law: Euthanasia Society Will Ask New Hampshire Act over Dr. Sander Murder Case," *New York Times*, January 3, 1950, ProQuest Historical Newspapers, *The New York Times 1851–2007*, p. 7.

6. "New Hampshire's Case," *New York Times*, January 8, 1950, ProQuest Historical Newspapers, *The New York Times 1851–2007*, p. 128.

7. *Id.*

8. "Not Guilty," *New York Times*, February 12, 1950, ProQuest Historical Newspapers, *The New York Times 1851–2007*, p. 128.

9. Blair, William M., "Euthanasia Issue in Medical Session: General Practitioners' Heads Fear Battle, Call It Moral, Not Scientific, Question," *New York Times*, February 21, 1950, ProQuest Historical Newspapers, *The New York Times 1851–2007*, p. 33.

10. *Id.*

11. "Physician Enables Incurables to Die: Provides Patients with Drug, Warning It Is Lethal, He Tells Euthanasia Meeting, Practice Common He Says," *New York Times*, January 18, 1950.

12. *Id.*

13. *Id.*

14. Porter, Russell, "Mercy Death Trial Begins Tomorrow: names of Prospective Jurors Kept Secret—Dr. Sanders [*sic*] Faces Prison, Hanging," Special to the *New York Times*, February 19, 1950, ProQuest Historical Newspapers, *The New York Times 1851–2007*, p. 12.

15. *Id.*

16. *Id.*

17. *Id.*

18. *Id.*

19. *Id.* For further discussion of the Kevorkian cases, see Chapter 5.

20. *Id.*

21. *Id.*

22. Porter, Russell, "Sander Witnesses Chorus Affection: 23 Men and Women Testify to Love and Admiration for 'Mercy Death' Physician, Colleagues Defend Him, Dr. Snay Declares He Found Mrs. Borroto Dead before an Injection in Vein," Special to the *New York Times*, March 4, 1950, ProQuest Historical Newspapers, *The New York Times 1851–2007*, p. 30.

23. *Id.*

24. *Id.*

25. "Textual Excerpts from Testimony of Dr. Hermann Sander at his Trial," Special to the *New York Times*, March 7, 1950, ProQuest Historical Newspapers, *The New York Times 1851–2007*, p. 19.

26. Porter, Russell, "Sander Jury Visits the Scene of Death: Completed with All 13 Men, It Goes to Hospital, Site of Alleged 'Mercy Killing,' 70 Veniremen Question, Court Lifts Ban on Queries, Showing Religion Was Not Specifically Mentioned," Special to the *New York Times*, February 23, 1950, ProQuest Historical Newspapers, *The New York Times 1851–2007*, p. 26.

27. Porter, Russell, "Pastor Urges Gifts to Dr. Sander Fund: Curice, of Home Town, Talks on 'Mercy,' Says Religion Is No Only Keeping in Law," Special to the *New York Times*, March 6, 1950, ProQuest Historical Newspapers, *The New York Times 1851–2007*, p. 13.

28. *Id.*

29. *Id.*

30. "Group Will Seek Mercy-Death Law," *op. cit.*, Jan. 3, 1950.

31. "The Nation," "Medical Group Deals New Blow to Sander," *New York Times*, March 12, 1950, ProQuest Historical Newspapers, *The New York Times 1851–2007*, p. E2.

32. *Id.*

33. "Medical Group Deals New Blow to Sander," *New York Times*, April 21, 1950, ProQuest Historical Newspapers, *The New York Times 1851–2007*, p. 33.

34. Porter, Russell, "State Body Defers to Local on Sander; Medical Society Meets 3 Hours, Takes No Action on Doctor Acquitted of Murder, County Group Has Say," Special to the *New York Times*, March 13, 1950, ProQuest Historical Newspapers, *The New York Times 1851–2007*, p. 20.

35. *Id.*

36. *Id.*

37. "Sander's Medical License Restored and He Gets a Patient in 10 Minutes," special to the *New York Times*, March 13, 1950, ProQuest Historical Newspapers, *The New York Times 1851–2007*, p. 31.

38. *Id.*

39. Oral Evidence, House of Lords Select Committee on Medical Ethics, 1993–1994.

40. Porter, March 10, 1950, *op. cit.*

41. This 1985 book came out nearly three decades after the original trial.

42. Devlin, *op. cit.*, 219.

43. This will be developed further in Chapter 6.

44. Devlin, 1–3.

45. *Id.*

46. *Id.*, 4.

47. *Id.*, 5–6.

48. *Id.*, 6.

49. *Id.*, 7.

50. *Id.*, 7.

51. Monk, William, *Euthanasia or Medical Treatment in Aid of an Easy Death* (London: 1887), 4–5; quoted in, N. D. A. Kemp, *"Merciful Release": The History of the British Euthanasia Movement* (Manchester: Manchester University Press, 2002), 41.

52. Pappas, Demetra M., "Recent Historical Perspectives Regarding Medical Euthanasia and Physician Assisted Suicide," *British Medical Bulletin* 52, no. 2 (1996).

53. Devlin, *op. cit.*, 171.

54. *Id.*

55. *Id.*

56. *Id.*

57. *Id.*, 171–72.

58. *Id.*, 172.

59. Pappas, *op. cit.*

60. *Regina v. John Bodkin Adams*, Transcript of the Shorthand Notes of Geo. Walpole & Co., (Official Shorthand Writers to the Central Criminal Court), "Seventeenth Day," "Summing Up," Tuesday, April 9, 1957, 18.

61. *Id.*, 19.

62. *Id.*, unnumbered covering page.

63. *Id.*, 19.

64. Devlin, *op. cit.*, 231.

65. Williams, *The Sanctity of Life and the Criminal Law* (New York: Alfred A. Knopf, Inc., 1957). In the book, Williams also criticized Christian opposition to contraception, artificial insemination, sterilization, abortion, and suicide. The last of these had an impact upon the revision of law regarding suicide in the early 1960s and will be discussed in that chapter.

66. Kamisar, "Some Non-Religious Views against Proposed 'Mercy-Killing' Legislation," 42 *Minn. L. Rev.* 969 (1958).

67. St. John-Stevas, *Life, Death and the Law: Law and Christian Morals in England and the United States* (Bloomington: Indiana University Press, 1961).

68. *Id.*, 262.

69. Kamisar, *op. cit.*, 70.

70. *Id.*, 71.

71. Williams, *op. cit.*, 313.

72. Kamisar, *op. cit.*, 987.

73. St. John-Stevas, *op. cit.*, 273.

74. Pappas, *op. cit.*

75. Williams, *op. cit.*, 342.

76. Kamisar, *op. cit.*, 120.

77. Williams, *op. cit.*, 314.

78. *Id.*, 322.

79. *Id.*, 319.

80. *Id.*, 321.

81. *Id.*, 381.

82. *Id.*, 326.

83. *Id.*, 325.

84. *Id.*, 340.

85. *Id.*, 342.

86. *Id.*, 342–43.

87. *Id.*, 345.

88. *Id.*, 342.

89. Kamisar, *op. cit.*, 1005–10.

90. St. John-Stevas, *op. cit.*, 265.

91. Williams, *op. cit.*, 315–16. There are many who would, in fact, disagree with this, and note that in cases of ethnic cleansing and genocide, soldiers are indeed examples of the slippery slope.

92. Kamisar, *op. cit.*, 973.

93. St. John-Stevas, *op. cit.*, 275.

4

The 1960s–1980s: Decriminalizing Suicide and Non-Voluntary Euthanasia of Those in Persistent Vegetative States

INTRODUCTION

After the 1950s, a long-held view—that suicide was a crime against the state—was effectively negated. In fact, in the United Kingdom, this was universally done in England and Wales by the *Suicide Act 1961*, although the same act criminalized assisting in a suicide. In the United States, this was an ongoing process from the 1960s onward, until no state had suicide as a crime. (However, a person who is suicidal could still be committed for his or her own protection, if they might be a danger to themselves or others.) Some of this legal change may perhaps be attributable to the debates of the 1950s.

During the same time that the technical (and mostly unenforced) crime of suicide was being revisited and revoked, a different—and medical—technology was causing a change in the view of the law regarding prolonging human life. This had a bifurcated cause. First, with the defeat of acute disease, and with degenerative illness coming into prominence at the end of the nineteenth century and the beginning of the twentieth century, cancer had fueled the modern euthanasia debate. Surgical developments led to the possibility of people being able to live longer. However, there was also sweeping social and medical change due to the increasing prominence of degenerative and late-onset illnesses, given the fact that life expectancy doubled from the norm of 40 years in 1851. In this process, the way in which people began living with illness from which they would have previously been dying also changed.

As a combination of these factors, much of the groundwork for the modern assisted-suicide debates of the 1990s was set. Technically, the Act announced itself as "the *Suicide Act 1961*."

THE *SUICIDE ACT* 1961

The theory of Anglo-American criminal law regarding both suicide and assisting in a suicide was subject to change in and after the 1960s. The *Suicide Act 1961* (of England and Wales) is a paradigmatic referent, allowing two essential matters that led to the medical euthanasia and assisted-suicide debates of the 1990s becoming readily apparent. First, as a general purpose, the law was intended to remove penalties for the survivors of a suicidant, who was "deemed to have taken life that was the prerogative of God and, therefore, not entitled to eternal salvation . . . not permitted burial in consecrated ground."[1] Traditionally, since medieval times in Western Europe, the family of suicidants also was penalized, including by forfeiture of property (which would have gone to the family and heirs of the deceased) to the state.

Thus, the purpose of Section 1 of the *Suicide Act* stated that "[t]he rule of law whereby it is a crime for a person to commit suicide is hereby abrogated,"[2] was a recognition of the fact that forfeiture to the Crown was actually a punishment on the family of the person who committed suicide, rather than upon that person. There was, as a practical matter, no way in which to prosecute a person who had committed suicide, or effectively engaged in their own homicide, with themselves as both perpetrator and victim of the crime.

However, this was not to say that the state viewed itself as having any the less of an interest in protecting the human life of its subjects. In fact, under the *Suicide Act 1961*, Parliament created crimes pertaining to promoting, soliciting, or aiding in a suicide or a suicide attempt. Section 2, regarding "Criminal liability for complicity in another's suicide," created a felony, in specific stating that "a person who aids, abets, counsels or procures the suicide of another, or an attempt by another to commit suicide, shall be liable on conviction on indictment to imprisonment for a term not exceeding fourteen years."[3]

As an initial matter, a reader considering this statute in the 2000s would look at it with the eye of one who has heard of assisted suicide, and would almost certainly be familiar with who Jack Kevorkian was and his activities in Michigan (so much so that the phrase "to Kevorkian" became one of common parlance). There had been no physician-assisted suicide cases on either side of the Atlantic Ocean. Indeed, the first uses of the phrase physician-assisted suicide, which were to emanate from the 1991 prosecutions of Dr. Timothy Quill in New York and Dr. Jack Kevorkian in

Michigan, were some 30 years away. Thus, the *Suicide Act 1961* was not created with regard to the medical wishes of terminally ill patients or patients suffering from degenerative illness.

The subsection is replete with self-described legislative intent. First, while decriminalizing suicide itself, by affirmatively criminalizing assisting, advising, or soliciting another person's suicide, Parliament indicated that it was not approving suicide or self-homicide. Rather, Parliament decreed that a participant would effectively be treated as aiding and abetting in a homicide. While later Anglo-American jurisdictions, such as Michigan, would enlarge this to include (in legislation criminalizing assisted suicide) and to specify providing the means or opportunity by which someone could commit suicide, this provision of Section 2 was clear in articulating that criminal liability would not only be the result, but that it would be felonious in nature. This is indicated by the possibility of conviction by indictment and by the lengthy sentence of 14 years.

At this point, it is interesting to note that while the *Suicide Act 1961* predated the criminal investigations of New York's Dr. Quill and Michigan's Dr. Kevorkian, the statutes that regarded their respective assisted suicides each would have provided for shorter sentences than the English enactment. In 1991, when Quill was investigated, Monroe County district attorney Howard Relin told the *New York Times* that "people convicted of aiding in a suicide can be sentenced to up to four years in prison."[4] This would be consistent with the law of manslaughter in the second degree, a class C felony, which has in its legislative history notes a creation date of 1965. Under that law, "[a] person is guilty of manslaughter in the second degree if . . . he intentionally causes or aids another person to commit suicide."[5] Whereas Quill was subjected to investigation further to a preexisting law (that was not created regarding medically assisted suicides, and which Quill went on to unsuccessfully challenge in a civil action that went to the U.S. Supreme Court in 1997), Kevorkian's activities in assisting suicides were the specific target of the Michigan legislature, which nevertheless provided for a lesser sentence of four years,[6] some 10 years less than that to which England was subjecting potential aiders and abettors of suicides (albeit not with a medical or treatment intention). At the time of the enactment of the *Suicide Act 1961*, medical assistance in suicide was not under consideration, hence perhaps the lengthy view of sentencing structure.

Parliament also decreed that "if on the trial of an indictment for murder or manslaughter it is proved that the accused aided, abetted, counseled or procured the suicide of the person in question, the jury may find him guilty of that offence."[7] It would be highly unusual for activities as a lesser included offense of murder to meet the elements of the new crime of aiding in a suicide. Assisting in a suicide is not an element-by-element lesser included offense of murder (although in the 2000s, one Georgia woman would be allowed to plead guilty to assisted suicide, after she shot her two sons to death as they lay in a nursing home, victims of late-state Huntington's disease). It would be equally unanticipated for the circumstances of general manslaughter (such as extreme emotional defense or provocation) to be found to have a lesser included component of assisting in a suicide, which seems almost certain to involve intentionality on the part of the assistant. Rather, it would appear that the purpose of this provision was to provide an alternative to jury nullification, or to offer a compromise verdict for a jury.

KAREN ANN QUINLAN: THE SEMINAL AMERICAN CASE REGARDING PERSISTENT VEGETATIVE STATE

If the *Suicide Act 1961* (and its American state cousins) demonstrated the sea change relating to the decriminalization of suicide, then the seminal 1976 New Jersey case of *In re Quinlan* set the stage for another sea change—that regarding the withdrawal of life support in cases of patients in a persistent vegetative state, in the best interests of the patient. Together, the decriminalization of suicide and the establishment of law regarding the withdrawal of treatment for those in persistent vegetative states would set the stage both for civil right-to-die cases and review of the law regarding physicians who engaged in assisted suicide.

On April 15, 1975, Karen Ann Quinlan was rushed to a hospital, after friends called emergency services. While it is theorized that a combination of liquor and pills caused her to pass out, in subsequent court proceedings, the Chancery Court first hearing the case ruled that "the precise events leading up to her admission to Newton Memorial Hospital are unclear."[8] What was very clear was that the 21-year-old had stopped breathing twice before arriving at the hospital, and given CPR first by friends, then by a police respirator. Medical tests showed a variety of drugs in her system, along with drink mixers. She would never regain consciousness.

Quinlan's father, a religious Catholic, sought a declaratory judgment and guardianship so as to be able to seek withdrawal of treatment. The state of New Jersey, the county, physicians, and the hospital were defendants in the action. Collectively, they contended that, "no constitutional right to die exists [, . . .] a compelling state interest in favor of preserving human life . . . [and] the court should not authorize termination of the respirator—that to do so would be homicide and an act of euthanasia."[9]

In other words, the court was presented with a question as to whether withdrawal of treatment, and allowing a person to die from the underlying condition (or a "natural death"), would constitute homicide as being an act of what would be tantamount to non-voluntary medical euthanasia. This question, in and of itself, implicated a variety of issues. First, the fact that Karen Ann Quinlan was in a persistent vegetative state, from which she would never be able to recover, meant that there was no opportunity for her to make a statement as to what her wishes would have been under these circumstances. Her lack of capacity rendered any acts that would hasten her death, whether by omission or by commission, not voluntary in nature. Second, but for the advances in medical technology, she would not have had the life support that her parents were seeking to discontinue. That is to say, had she not been hooked up to a respirator in the first place and had "extraordinary medical procedures [then] sustaining Karen's vital processes and hence her life,"[10] she would have died from natural causes within days (if not hours or minutes) of the originating events that led to her loss of consciousness.

While Karen Ann Quinlan's father sought to remove his daughter from artificial life support, her primary care physician and the hospital refused, and state prosecuting authorities contended that intervening to remove life support once started could constitute homicide. There was no question that the Quinlans were a close-knit family and that Karen Ann remained an important member of the family after she moved out as a young woman, nor was there any question that any proceedings brought by Joseph Quinlan were in accordance with love for his child, and without any intention of benefit to himself:

> The character and general suitability of Joseph Quinlan as a guardian could not be doubted. The record bespeaks the high degree of familial love which pervaded the home of Joseph Quinlan and reached out fully to embrace Karen.[11]

In these and other words, the court depicted family love and unity. While this was the first landmark case to lead to right-to-die cases and legislation, as well as the assisted-suicide cases (both civil litigation and criminal prosecution) and legislation, there are two predictive notes here. First, the level of family unity discussed may have influenced Jack Kevorkian in his 1990s assisted-suicide and euthanasia activities, to the extent that there was testimony at his trials (and at his sentencing, following the 1999 conviction for his medical euthanasia of Tom Youk), that, in addition to seeking consent from the patient (or victim or client, depending upon one's political or legal stance), he sought unity of consent from the families of his decedents. In a completely different vein, during the 2000s, another young woman in a persistent vegetative state, Terri Schiavo, was the subject of family acrimony and disunity between her husband Michael (who wanted to discontinue life support) and her parents (who wanted life support continued), and who litigated the matter through the judicial system in what was nothing less than a family feud. These two series of events would show two poster children—Jack Kevorkian was unquestionably the poster child of assisted suicide and medical euthanasia in the 1990s (and thereafter), whereas Terri Schiavo became a poster child for the fate of those in a persistent vegetative state (though not of physician-assisted suicide or medical euthanasia). Unlike Terri Schiavo, who was shown in photographs and in video in major media outlets while in a persistent vegetative state, Karen Ann Quinlan was generally shown as an attractive young woman with long brown hair and a steady gaze, all prior to her decline.

The Quinlans, in addition to being a strong family, were also very Catholic. The records suggest that the family was pro-life in its views and that this had an impact upon its decision-making process. Joseph Quinlan consulted both with his parish priest and with the hospital chaplain before seeking to have Karen Ann's life support removed. Indeed, the Supreme Court of New Jersey noted that Joseph Quinlan was:

> deeply religious, imbued with a morality so sensitive that months of tortured indecision preceded his belated conclusion . . . to seek the termination of life-supportive measures sustaining Karen.[12]

As to Quinlan's religious process in the decision making, the court noted him to be "a communicant of the Roman Catholic Church, as were other family members, he first sought solace in private prayer."[13] Moreover, Quinlan:

> consulted with his parish priest and . . . Catholic Chaplain of St. Clare's Hospital. He would not, he testified, have sought termination if that act were to be morally wrong or in conflict with the tenets of the religion he so profoundly respects.[14]

The Supreme Court of New Jersey, while observing that "it is not usual for matters of religious dogma or concepts to enter a civil litigation (except as they may bear upon constitutional right or sometimes familial matters),"[15] determined that because matters of faith were so intertwined with the Quinlans, they were properly admitted and considered by the court of first impression, and were appropriate for them to consider as part of the case. The Supreme Court noted that it had:

> no reason to believe that [terminating Karen Ann Quinlan's life support] would be at all discordant with the whole of Judeo-Christian tradition, considering its central respect and reverence for the sanctity of human life.[16]

Indeed, Bishop Lawrence B. Casey, of the New Jersey Catholic Conference, offered views that the New Jersey Supreme Court determined "validated the decision of Joseph Quinlan."[17] While perhaps unexpected, this too proved to be predictive in another matter, in a surprising way. Merian Frederick, for whose assisted suicide Jack Kevorkian was tried and acquitted in 1996, not only solicited the counsel of her minister, Rev. Kenneth Phifer, but was successful in her request to ask the minister to be present on October 22, 1993, when she died with Kevorkian's assistance. In short, while some might anticipate that men (and women) of the cloth might be unrelentingly opposed to hastened death, there are circumstances (usually regarding removal of life support, rather than presence or support of physician-assisted suicide) when those with religious commitment nonetheless are neutral (do not oppose) or even supportive of hastened death.

By the time the Supreme Court of New Jersey heard the Quinlan case, Karen Ann was "described as emaciated, having suffered a weight loss of at least 40 pounds [from a normal adult weight]."[18] She was also described as "undergoing a continuing deteriorative process. . . . [with fetal-like] posture, extreme flexion-rigidity of the arms legs and related muscles and her joints."[19] The decline from vibrant, fun-loving, life-affirming 21-year-old to this state was accepted by all to be heartwrenching.

Nevertheless, while remaining in an intensive care unit at St. Clare's Hospital, she was described as receiving 24-hour care by a team of four nurses characterized, as was the medical attention, as "excellent."[20] There was equally no question that the "excellent medical and nursing care so far has been able to ward off the constant threat of infection, to which she is peculiarly susceptible because of the respirator, the tracheal tube and other incidents in care of her vulnerable condition."[21]

However, there was also no question that "no physician risked the opinion that she could live more than a year and indeed she may die much earlier."[22] The New Jersey Supreme Court concluded that "her life accordingly is sustained by the respirator and tubal feeding, and the removal from the respirator would cause her death soon, although the time cannot be stated with more precision."[23] This 1976 phrasing introduced a construction of time to what would develop in the movement for assisted-suicide legislation and litigation—a year at the outside became a dateline for what would be considered "terminal" illness. That politicalization of time and language would not prove to be *de minimis*—after all, assisted-suicide legislation requiring a terminal illness of life that would otherwise end by underlying illness in one year or less would, in the 1990s and thereafter, leave out most with Alzheimer's disease, Huntington's disease, multiple sclerosis, Lou Gehrig's disease (ALS) and even late-stage AIDS (leaving aside for a moment the issue that the terminal stages of these illnesses might be at times when lucidity and capacity for voluntariness might come into question).

During the 1960s, sociologists Barney G. Glaser and Anselm Straus wrote a trio of books considering the end of life as experienced by patients at the end of life in hospitals, in which they considered hospitals, nursing, and family issues in the context of what came to be known as the "death trajectory," and indeed similarly subtitling their book, *Anguish: A Case History of a Dying Trajectory.*[24] The traditional natural trajectory, for example, unpunctuated by accident or acute illness or death in childbirth, would be for a person of advanced years to cease eating and drinking and to slip away, perhaps of pneumonia, which was also known colloquially as the old man's friend. Glaser and Strauss were focused upon a woman who was enduring a lengthy death trajectory of cancer, riddled with pain and suffering.

The *Quinlan* court, however, was dealing in *terra nova*—rather than dealing with a death trajectory from cancer or another degenerative illness, the court found itself faced with the question of what the definition of death itself was. In so considering, the court ruminated that:

> From ancient times down to the recent past [i.e., recent to the 1970s] it was clear that, when the respiration and heart stopped, the brain would die in a few minutes, so the obvious criterion of no heart beat as synonymous with death was sufficiently accurate. In those times, the heart was considered to be the central organ of the body; it is not surprising that its failure marked the onset of death. This is no longer valid when modern resuscitative and supportive measures are used. These improved activities can now restore "life" as judged by the ancient standards of persistent respiration and continuing heart beat. This can be the case even when there is not the remotest possibility of an individual recovering consciousness following massive brain damage.[25]

An ad hoc committee of the Harvard Medical School (which included 10 physicians, a historian, an attorney, and a theologian) issued standards defining "irreversible coma," which "included absence of response to pain or other stimuli, pupilary reflexes, corneal, pharyngeal and other reflexes, blood pressures, spontaneous respiration, as well as 'flat' or isoelectric electroencephalograms and the like, with all tests repeated . . . [to demonstrate] brain death."[26] The *Quinlan* court noted that one issue was that the "consensus of medical testimony . . . [was] that Karen, for all her disability,"[27] did not meet all of the criteria of the ad hoc committee. Nevertheless, the New Jersey Supreme Court determined that "competent medical testimony [had] established that Karen Ann Quinlan [had] no reasonable hope of recovery from her comatose state by the use of any [then] available medical procedures."[28]

Furthermore, the New Jersey Supreme Court recognized the potential for criminal liability (as well as civil liability for malpractice) that could await doctors who terminated life support, stating that "there must be a way to free physicians, in the pursuit of their healing vocation, from possible contamination by self-interest or self-protection concerns which would inhibit their independent medical judgments for the well-being of their patients."[29] Citing an article by Dr. Karen Teel, the court noted that:

[p]hysicians, by virtue of their responsibility for medical judgments are, partly by choice and partly by default, charged with the responsibility of making ethical judgments which we [doctors] are sometimes ill equipped to make. We [doctors] are not always morally and legally authorized to make them. The physician is thereby assuming a civil and criminal liability that, as often as not, he [*sic*] does not even realize as a factor in his decision. There is little or no dialogue in this whole process. The physician assumes that his judgment is called for and, in good faith, he [*sic*] acts.[30]

One issue in the *Quinlan* case was that while the medical obligation was related to standards and practice in the profession, the emerging technology provided for the prolonging of the death trajectory, or even its suspension, without providing for standards. Indeed, the very definition of a "persistent vegetative state," by Dr. Fred Plum and Dr. Bryan Jennett, was itself emerging during this time period.[31] Thus, the question of how to treat and what the legal parameters of life and death themselves continued to emerge after their 1972 seminal article. The uncertainty of whether one in a persistent vegetative state should be classified as alive or dead presented the cascading questions of whether to terminate life support could be viewed as non-voluntary euthanasia, and in turn homicide.

The New Jersey Supreme Court acknowledged the thorniness of these issues in its holding, stating that it "thus arrive[d] at the formulation of the declaratory relief which we have concluded is appropriate to this case."[32] In essence, the very wording of this preamble to the court's final conclusions and decision suggested that it was on uncertain ground, as well as *terra nova*. While the 1976 case would be the standard against which other cases of how to handle requests to terminate life support would be handled, on the face of it, the court restricted its ruling to the limited facts and case before it, leaving open the possibility that it might not be a sweeping precedent. With focus upon Karen's "continuing deterioration . . . [and the fact that] she is now even more fragile and nearer to death . . . and that there is no reasonable possibility of Karen's ever emerging from the present comatose condition to a cognitive, sapient state,"[33] the life support could be discontinued, with Joseph Quinlan as Karen's guardian, and that "said action shall be without any civil or criminal liability therefore on the part of any participant, whether guardian, physician, hospital or others."[34]

In addition to absolving doctors, hospital, and guardian from criminal risk, the New Jersey Supreme Court made a statement that would, some

30 years later, stand in contrast to the family feud that resounded in the wake of Terri Schiavo's husband and parents litigating who should be her guardian—either the husband who would have sought to terminate support or the parents who would have sought to continue it. The New Jersey Supreme Court focused upon the unity of the family, as well as the medical providers, specifically stating:

> We repeat for the sake of emphasis and clarity that upon the concurrence of the guardian and family of Karen, should the responsible attending physicians conclude that there is no reasonable possibility of Karen's ever emerging from her present comatose condition to a cognitive, sapient state and that the life-support apparatus now being administered to Karen should be discontinued, they shall consult with the hospital "Ethics Committee" or like body of the institution in which Karen is then hospitalized . . . and the life-support system may be withdrawn and said action shall be without any civil or criminal liability therefore, on the part of any participant, whether guardian, physician, hospital or others.[35]

Thus the court reiterated that no physician (or medical provider) would be subject to medical malpractice, criminal prosecution, or other liability regarding removing Karen Ann Quinlan from life support. The court then went one step further, recognizing the precedential value of its decision, notwithstanding the earlier declaration that the judges were deciding the case on the facts before it; this the court did with the simple statement that "by the above ruling, we do not intend to be understood as implying that a proceeding for judicial declaratory relief is necessarily required for the implementation of comparable decisions in the field of medical practice."[36] Thus it was that the seminal case absolving American doctors from prosecution for actions tantamount to non-voluntary euthanasia closed, with litigation and legislation to follow in Anglo-American jurisdictions.

A postscript to the story of Karen Ann Quinlan is that while her life support was disconnected in 1976, she did not die until June 11, 1985. This is because she was able to breathe on her own after life support was disconnected, and because nutrition and hydration were never withdrawn, so that she died years later of pneumonia ("the old man's friend").

THE SUPREME COURT DECISION IN *CRUZAN V. DIRECTOR, MISSOURI DEPARTMENT OF HEALTH*

Nancy Cruzan, like Karen Ann Quinlan, was a young woman in good health until a catastrophic event left her in a persistent vegetative state. After a car accident on January 11, 1983, Cruzan was thrown from her car and landed face down in a ditch filled with water. Although the state was financing her care in a state hospital, her parents sought to terminate her artificial nutrition and hydration, given her vegetative state. The Missouri hospital refused to do this, and hence her parents brought a civil action, in which they succeeded in securing an order for termination of life-sustaining treatment. This the Missouri Supreme Court reversed, on the grounds that there was no clear and convincing evidence that this was what Nancy would have wanted under these circumstances, although Nancy Cruzan had told her housemate that she would not want to live if anything happened to her and that "if sick or injured she would not want to continue her life unless she could live 'halfway normally.' "[37]

One noteworthy difference in the Cruzan case is that, unlike the Quinlan family, the Cruzans were seeking termination of artificial nutrition and hydration. The Cruzans, seeking to rely on the doctrine of substituted judgment to permit them to choose termination of life support, were seeking to what they believed to be their daughter's wishes. The U.S. Supreme Court majority stated that "no doubt is engendered by anything in this record but that Nancy Cruzan's mother and father are loving and caring parents [and i]f the State were required by the United States Constitution to repose a right of 'substituted judgment' with anyone, the Cruzans would surely qualify."[38] This, like the Quinlan case, would later stand in contrast to the Schiavo case, where both the husband and the parents were questioning each other's motives and alleging financial motivations to gain access to her trust fund. The *Cruzan* court went on to say that "we do not think the Due Process Clause requires the State to repose judgment on these matters with anyone but the patient herself."[39] Subsequent to the U.S. Supreme Court case, the Cruzans were able to bring additional proof to the Missouri court and to secure an order allowing them to discontinue nutrition and hydration. Cruzan died 11 days later, on December 26, 1990.

It was, however, the dissent of Justice Scalia that perhaps contributed the most commentary to the soon-to-be developed assisted-suicide debate. Scalia noted that:

> American law has always accorded the State the power to prevent, by force if necessary—suicide—including suicide by refusing to take appropriate measures necessary to preserve one's life; that the point at which life becomes "worthless" and the point at which the means necessary to preserve it become "extraordinary" or "inappropriate" are neither set forth in the Constitution nor known to the nine Justices of this Court any better than they are known to nine people picked at random from the Kansas City telephone directory; and hence, that even when it is demonstrated by clear and convincing evidence that a patient no longer wishes certain measures to be taken to preserve her life, it is up to the citizens of Missouri to decide, through their elected representatives, whether that wish will be honored.[40]

In a sense, this dissent predicted both that the U.S. Supreme Court would be disinclined to rule that there was a federal due process or an equal protection right to assisted suicide in 1997, and also that the court would be disinclined to overturn the Oregon Death with Dignity Act in 2005. Thus, Scalia in fact acknowledged that there were "agonizing questions that are presented by the constantly increasing power of science to keep the human body alive for longer than any reasonable person would want to inhabit it,"[41] but challenged the legal methodology.

As Scalia pointed out, suicide had been a crime at common law in England, which was "defined as one who 'deliberately puts an end to his own existence or commits any unlawful malicious act, the consequence of which is his own death' was criminally liable."[42] Moreover, Scalia pointed out that when "the States abolished the penalties imposed by the common law (i.e., forfeiture and ignominious burial), they did so to spare the innocent family, and not to legitimize the act."[43]

CONCLUSION

As an irony, the cases of Nancy Cruzan and Karen Ann Quinlan were more about the right not to be treated, the right to refuse treatment, or the mirror image of informed consent. That said, the time period from the 1960s until the 1990s was one in which two critical sea changes took place. First, suicide was definitively decriminalized, although assisting in a suicide was deemed a crime. Second, a body of law was developing in which a refusal to endure further treatment, and by which guardians could enforce those rights, became entrenched in American society. The next decade would culminate in a tsunami of legislation, civil litigation, and criminal prosecutions in which assisted suicide was both criminalized (as in Michigan) and decriminalized (as in Oregon).

NOTES

1. Howarth, Glennys, *Death and Dying: A Sociological Introduction* (Cambridge: Polity Press, 2007), 219 (*citing* Gittens, 1984).
2. *Suicide Act 1961, Section 1.*
3. *Suicide Act 1961, Section 2 (1).*
4. Altman, Laurence K., "Doctor Says He Gave Patient Drug to Help Her Commit Suicide," *New York Times*, March 7, 1991.
5. *New York Penal Law, Section 125.15 (3).*
6. *Michigan Code Archive Directory, Crimes and Offenses, Assistance to Suicide, Section 752.1027* (1993).

7. *Suicide Act 1961, Section 2 (2).*
8. *In re Quinlan*, 173 N.J. 227, 237 (1975).
9. *Id.*, 251.
10. *In re Quinlan*, 70 N.J. 10 (Sup. Ct. N.J., 1976).
11. *Id.*, 30.
12. *Id.*
13. *Id.*
14. *Id.*
15. *Id.*
16. *Id.*
17. *Id.*, 32.
18. *Id.*, 26.
19. *Id.*
20. *Id.*
21. *Id.*
22. *Id.*
23. *Id.*
24. Strauss, Anselm L. and Glaser, Barney G., *Anguish: A Case History of a Dying Trajectory* (San Francisco: University of California Medical Center, 1970).
25. *In re Quinlan*, 70 N.J. 10, 27.
26. *Id., citing* 205 *J.A.M.A.* 337, 339 (1968).
27. *Id.*
28. *Id.*, 32.
29. *Id.*, 49.
30. *Id., citing* Karen Teel, "The Physician's Dilemma: A Doctor's View: What the Law Should Be," *Baylor Law Review.*
31. Jennett, Bryan and Fred Plum, "Persistent Vegetative State after Brain Damage: A Syndrome in Search of a Name," *Lancet* 299, no. 7753 (1972): 734–37.
32. *In re Quinlan*, 70 N.J. 10, 54 (Sup. Ct. N.J., 1976).
33. *Id.*
34. *Id.*
35. *Id.*, 55.
36. *Id.*
37. *Cruzan v. Harmon*, 760 S.W.2d 408, 411 (1988).
38. *Cruzan v. Director, MDH*, 497 U.S. 261, 286 (1990).
39. *Id.*
40. *Id.*, 293.
41. *Id.*, 292.
42. *Id.*, 293.
43. *Id.*, 294.

5

The 1990s and Jack "Dr. Death" Kevorkian: From Physician-Assisted Suicide to Medical Euthanasia

INTRODUCTION

Often, controversy surrounds or emanates from a specific case, such as the U.S. Supreme Court's 1973 decision in *Roe v. Wade*,[1] the "firebrand" in the pro-life/pro-choice abortion controversy.[2] For most of the American public, the lightning rod in the euthanasia and assisted-suicide controversy was Dr. Jack Kevorkian. He admitted to assisting (and in some cases, causing) the deaths of over 130 people between 1991 and 1998. This is in contrast to New Hampshire's Hermann Sander, who was tried in 1950 to acquittal of a patient who was deemed to be "already dead." Kevorkian was also unlike England's John Bodkin Adams, who was tried once and acquitted in 1957 of one patient who was deemed to have died due to the "double effect" of drugs intended to relieve suffering. Jack Kevorkian was tried not once, not twice, but on five separate occasions, regarding seven unique cases variously of euthanasia (as a murder charge) and assisted suicide, prior to his final conviction in 1999.

Kevorkian made news so much as to become both a media darling and a folk devil during the 1990s. With so many cases, these provide historical contexts of different theories of medical cause of death, criminal intent (or lack thereof), and to provoke prosecutorial, judicial, administrative, and legislative responses in a variety of contexts. In short, he became a poster child for all that might be argued both in favor of, and opposition to, euthanasia and assisted suicide during the 1990s.

The mere mention of Jack Kevorkian, the self-styled "Dr. Death" of Michigan, in the United States or another Anglo-American jurisdiction, is enough to get spirited discussion going on both sides of the euthanasia and assisted-suicide debate. Kevorkian, a former pathologist, became a one-person catalyst for this issue, notwithstanding the contemporaneous activities of other doctors, lawyers, and patients, who are discussed in other chapters. He assisted in suicides, performed euthanasia, "dropped off" (delivered) bodies at the local hospitals, made videotapes, and repeatedly went on television—all the while provoking, and even demanding, criminal prosecution by local authorities. Indeed, by the end of the twentieth century, his name had become synonymous with euthanasia in popular culture. HBO made a movie, *You Don't Know Jack*,[3] starring Al Pacino in an Emmy-winning turn in 2010, all made with Kevorkian's approval.[4] The tape-recorded euthanasia of his client/patient/victim, Tom Youk, was broadcast on *60 Minutes* in November 1998 and was largely instrumental in Kevorkian's conviction of murder in the second degree the following year.

He was brought to trial, variously, in 1994 (for assisting in a suicide), 1996 (for assisting in two suicides), again in 1996 (on charges of "open murder"), 1997 (for assisting in a suicide that most likely was a euthanasia by injection), and lastly, in 1999 (for murder). Four different chief prosecuting attorneys initiated these actions. These men ranged from pro-choice views (John O'Hair, the chief prosecuting attorney in the 1994 case) to pro-life views (Richard Thompson, who initiated the two double cases brought to trial in 1996), with the more moderate Ray Voet (in 1997) and the more practical David Gorcyca (in 1999). Juxtaposed with each other in and of themselves, they too demonstrated the historical debates of the 1990s in context.[5] This is because the chief prosecuting attorneys had a chance to shape not only the criminal prosecutions against Jack Kevorkian, but also the assisted-suicide and euthanasia debate of Michigan, which was broadcast and reported upon throughout the world. This also proved to be true of the judges who tried the Kevorkian cases. The same may be said of at least some of the jurors who deliberated and determined verdicts in the Kevorkian trials.

Unraveling all of these is difficult, especially when various cases were on hold even while Kevorkian engaged in other acts of hastened death. Thematic issues overlapped with chronological development. As a result, this chapter is organized in equal measure by participants in the criminal trials, in chronological order of events—weaving the fabric of the then-emerging debate in Michigan.

IN THE BEGINNING, 1990–1993: AN ASSISTED-DEATH PROTAGONIST, AN ANTAGONISTIC PROSECUTOR, AND AN ANTI-KEVORKIAN LEGISLATIVE ENACTMENT

That Jack Kevorkian was a licensed medical doctor made all the difference between a likely successful prosecution, criminal conviction, and prison sentence immediately after providing an "assisted suicide machine" to Janet Adkins to use in the back of his 1968 VW van on June 4, 1990. While Richard Thompson, the chief prosecuting attorney, brought a first-degree murder charge against the retired pathologist, he declined to appeal when the charges were dismissed six months later by Judge Gerald McNally. The basis of McNally's December 12, 1990, ruling rested largely upon the viewing of a videotape in which the Alzheimer's patient repeatedly told Kevorkian, "I don't want to go on."

In a society in which surgeries are routinely recorded as a matter of hospital protocol, this might equally be viewed as an effort to avoid civil malpractice (consent notwithstanding) or criminal liability. Either way, Kevorkian learned a valuable lesson, and began to appear in public and on television with would-be assisted suicidants, and also to videotape consent colloquies in which patients asked for him to help end their lives. On October 23, 1991, Kevorkian reported the deaths of Marjorie Wantz, who suffered from what the courts euphemized as "intense pain in the pelvic and vaginal area for twenty years,"[6] and Sherry Miller, a middle-aged woman who had been diagnosed with multiple sclerosis some 20 years earlier and was confined to a wheelchair.

Chief Prosecuting Attorney Thompson also noted that Kevorkian's timing was also acute—he observed that the "Wantz/Miller assisted suicides happened the week before the election [when California had a ballot measure] and the ME [Medical Examiner or Coroner] said the scene [of the Wantz/Miller Kevorkian assisted suicides] was bizarre, a cabin lit by candles" (Thompson interview, August 20, 1993). According to the Court of Appeals, Wantz was "hooked up to [Kevorkian's] 'assisted suicide machine,' which consisted of a board to which her arm was strapped . . . , a needle . . . and containers of various chemicals that could be released into the needle . . . into the blood stream."[7] Because Kevorkian's efforts to connect the assisted-suicide machine to Sherry Miller failed, "he left the cabin . . . procured a tank of carbon monoxide gas and a mask assembly . . . attached the mask to [her] face and instructed her"[8] how to use the gas valves, successfully, as she died of carbon monoxide poisoning.

Thompson said that he felt obliged to prosecute Kevorkian after Kevorkian's first assisted suicide. Once that case was dismissed, Thompson sought "clarification" from the legislature about the law regarding medical euthanasia and assisted suicide (Thompson interview, August 20, 1993). What he was seeking was to criminalize the conduct and to prosecute Kevorkian to a successful conclusion of conviction. In the meantime, on November 20, 1991, less than one month after the October 23, 1991, Wantz/Miller cases, the Michigan State Board of Medicine revoked Kevorkian's license to practice medicine, specifically on the grounds of his conduct in assisting in the deaths of Wantz and Miller.

Notwithstanding the loss of licensure, the fact that Kevorkian was a physician played no small role in Circuit Court Judge David Breck's initial decision to dismiss charges of murder that were (again) brought by Oakland County's chief prosecutor, Richard Thompson.[9] It is at this juncture that the Kevorkian cases had their true beginning, and it began raining litigation and legislation in Michigan.

Thompson, relying upon a Michigan Supreme Court decision from the 1920s, *People v. Roberts*,[10] charged Kevorkian with what was called common law "open murder"—these were the charges that were dismissed prior to trial, from which the prosecutor appealed and got the charges reinstated (and ultimately tried in 1996). The central reason *People v. Roberts* was affirmed was because that defendant had entered a plea of guilty, to the charge of murder in the first degree, in which he confessed to having mixed a poison (Paris green) with water and placed in wife Katie's reach to enable her to end her own life.

There are two important aspects of the *Roberts* case. First, Frank Roberts told the court in his plea colloquy (and in response to a direct inquiry from the court) that his wife had previously attempted suicide the summer before[11] and then knowingly drank the poison. This was in a way that would in today's parlance and *current* (but not then-existing) Michigan law be deemed assisted suicide. This is supported by the coroner who performed the autopsy—who knew Katie Roberts while she was alive, had seen her "about three or four months before her death . . . at her home where they lived . . . [and that] she was a bed patient [with] her body considerably wasted."

Second, the issue in the challenge in *Roberts* was whether a murder trial was required subsequent to an allocuted guilty plea, in which the defendant knowingly and voluntarily enters a plea of guilty and describes to the court the facts of causation (actus reas) and intent (mens rea) constituting the crime. Most pertinent in the *Roberts* case, a condition precedent to an allocuted guilty plea is a colloquy in which the defendant legally waives the right to a trial by jury, and in which a defendant is advised that a guilty plea has the same legal effect as a conviction after trial. The Michigan Supreme Court accordingly held that "there is no provision of the constitution which prevents a defendant from pleading guilty to the indictment instead of having a trial by jury. If he [Mr. Roberts] elects to plead guilty to the indictment the provision of the statute for determining the degree of the guilt for the purpose of fixing the punishment does not deprive him of any right of trial by jury."[12]

However, Judge Breck did not agree with the prosecutor's arguments about the applicability of *Roberts*, and in specific wrote, "[i]n this case, the defendant is a doctor who was acting in the course of a physician-patient relationship (and this Court so holds, based upon the lower court testimony)."[13]

Judge Breck continued that "this is a case of first impression at the Circuit Court level,"[14] noting that, "assisted suicide, whether physician assisted or not, is legal."[15] However, perhaps most controversial was Judge Breck's "conclusion that physician assisted suicide is not a crime in Michigan, even when the person's condition is not terminal."[16] The controversy ensued over the fact that the Michigan Court of Appeals, in *People v. Campbell*,[17] an appellate court (of lower stature than the Supreme Court) ruled that incitement to suicide was not a crime; however, in that case of a drunken lover's triangle, the defendant gave a gun to the decedent, in what the court ruled was not a present intention to incite a suicide. This was, aside from being subordinate to the *Roberts* case (which admittedly may not have been quite on point to Kevorkian, due to the plea matter), clearly distinct in the intent component of the Kevorkian cases (if only as a legal fiction). Judge Breck was, however, clearly correct in that the physician-assistance issue was presenting for the first time in Michigan as a result of Kevorkian's activities.

Thompson developed a series of issue-related arguments in the course of appealing from Judge Breck's decision (which hinted at the philosophy that Breck would have toward both Kevorkian and the euthanasia/assisted-suicide controversies when he ultimately had to try the Wantz/Miller cases in 1996, after higher courts reversed his decision). Thompson went further, talking about possibilities of resorting to hospice or anticipating palliative care, rather than assisted death. He argued, "the medical profession maybe is starting to understand [during the 1990s] if we can't cure them, we can care for them" (Thompson interview, August 20, 1993). He also alluded to the "slippery slopes" argument: "once you say there is such a thing as life not worth to be lived it opens up . . . slopes [*sic*] are coming to fruition in the Netherlands with . . . mentally defective babies."

Adults, with the legal (and medical) capacity to consent, are in an entirely different legal category than are underage people and mentally disabled infants who have no legal capacity to consent. As to "Dr. Death," Thompson said: "[W]hen I became prosecutor, this was not an issue we would talk about . . . [it] wasn't really an issue for us, it developed because Kevorkian was an Oakland County resident and he started killing people here in Oakland. But I want you to know we handle 20,000 felony cases a year in this office and we always maintain our focus on general responsibilities" (Thompson interview, August 20, 1993).

The procedural history regarding introduction of anti-assisted-suicide legislation shows that, in fact, it was indeed Kevorkian's first case, that of Janet Adkins, which prompted his quest for a criminal statute (written "black letter law") regarding assisted suicide, although he did not appeal the dismissal of that case. The first bills introduced before the Michigan House (HB 4038) and the Michigan Senate (SB 32), both entitled "Crimes; Homicide; Aided Suicides, Prohibit," were introduced on May 7, 1991[18]—that is to say, several months after Janet Adkins's death, and several months before the deaths of Marjorie Wantz and Sherry Miller. However, the final legal ban on assisting in a suicide was added to a preexisting bill that was originally intended to create a commission to study issues related to death and dying.[19] "There was no provision to criminalize the act of assisting a suicide in any part of the original HB 4501."[20]

On November 24, 1992, amendments were offered "adding a section 7 which created the felony of assisting a suicide."[21] These amendments were adopted, and a bill creating a Michigan Commission on Death and Dying and also criminalizing assisted suicide was passed on December 3, 1992, with an effective date of March 30, 1992. The provision of the new act, "Section 7" created a new crime of "criminal assistance to a suicide," making it a felony where a person did either of the following:

a. provides the physical means by which the other person attempts or commits suicide;
b. participates in a physical act by which the other person attempts or commits suicide.[22]

After Kevorkian assisted in the February 15, 1993 suicide of Hugh Gale, a 70-year-old man with emphysema and congestive heart failure, the Michigan legislature proposed a bill to give immediate effect to the assisted-suicide ban, which passed and took immediate effect on February 25, 1993. Michigan Right to Life was widely credited with mobilizing the immediate effect passage, as well as prompting prosecutor Carl Marlinga of Macomb County to initiate proceedings against Kevorkian.

It is well worth noting that after Public Act 270 (the original act) passed, and prior to Public Act 3 (the immediate effect act of the same enactment), Kevorkian did not desist, but rather escalated his activities. On December 16, 1992, Kevorkian assisted in the suicides of Marguerite Tate and Marcella Lawrence. On January 20, 1993, Kevorkian assisted in the death of Jack Miller. On February 4, 1993, Kevorkian assisted in the deaths of Stanley Ball and Mary Biernat. On February 11, 1993, the Michigan Senate approved a bill to provide for amendments to the assisted-suicide ban to exempt caregivers who did not have a medical license, but who did provide pain medication, further to a hospice program. On February 18, 1993, Kevorkian assisted in the deaths of Jonathon Grenz and Martha Ruwart. In Michigan parlance, the immediate effect passage of "the anti-Kevorkian law" had become an urgent matter, according to Senator Fred Dillingham (R-Fowlerville) (Dillingham interview, August 23, 1993). While the bill had a "sunset clause" to cease in its effect after 18 months (at which time, the newly created Michigan Commission on Death and Dying would presumably have made recommendations to the legislature to act upon), both civil challenges to the law and further Kevorkian activities and prosecutions took place with deliberate speed.

1994–1996: CIVIL LITIGATION AND CRIMINAL PROSECUTION

Less than one week after the assisted-suicide ban became law in Michigan, the American Civil Liberties Union filed a lawsuit in Wayne County (a relatively liberal part of the state, which has Detroit as its central city). Kevorkian was not a party to this civil lawsuit challenging the constitutionality of 1992 PA 270 and 1993 PA 3. Instead, the plaintiffs included seven health care professionals, two cancer patients, and "a friend of one of the lay persons," as described by the court.[23] In fact, the "friend" was a life partner (Teresa Hobbins/Marie DeFord interview, October 4, 1994), who was potentially at risk of criminal liability if she assisted Hobbins in any way. This risk was very real.

Actions by Marie DeFord would more likely have been within the ambit of the phrase "mercy killing" to distinguish from medical euthanasia and physician-assisted suicide. That is to say, these potential mercy killings are cases in which a husband shoots or suffocates a terminally ill wife, or a parent commits such acts on a child who is ill. Mercy killing has itself remained a political term. For example, in the Written Evidence of the House of Lords Select Committee on Medical Ethics, the Home Office included a table, "Offenses Recorded as Homicide Where Circumstance Was Coded as Mercy Killing: For a 10-Year Period of 1982–1991."[24] The table was illuminating of more than the number of deaths and arrests—of the 22 cases, only one was prosecuted as a murder, and in that case, the perpetrator was an "acquaintance male," who was convicted and received a life sentence; and the only other case involving an "acquaintance male" resulted in a two-year period of incarceration. There is a possibility that these "acquaintance males" were homosexual partners, and that the victims, who were aged 50 and 25, respectively, were persons with AIDS, arising at the same time as the advent of AIDS. Placing this in a historical context, a question emerges as to whether the criminal justice system marginalized people already marginalized by their sexual preference. In other words, the House of Lords Select Committee on Medical Evidence, which, like the Michigan Commission on Death and Dying, heard evidence and engaged in deliberations between 1993 and 1994, received a document attesting to the fact that while family members were treated lightly by courts in mercy-killing cases, receiving probation or light sentences, an "acquaintance" or same-sex life partner of a deceased was at risk of receiving a harsh sentences of a number of years.

On May 20, 1993, Judge Cynthia Stephens struck down the Michigan assisted-suicide ban, in specific holding that Hobbins and co-plaintiff Ken Shapiro demonstrated "a substantial likelihood of success on the merits on their claim of a liberty interest in ending their own lives. . . . and that the right of self determination [*sic*] rooted in the Fourteenth Amendment of the Federal Constitution (US Cont. [*sic*], Amend 10) and of the Michigan Constitution 1963 Mich Con [*sic*] Article I Section 17, includes the rights [*sic*] to choose to cease living" and for other reasons relating to procedural matters regarding dual purpose of one statute.[25]

Even as the Michigan Court of Appeals stayed (placed on hold) Judge Stephens's ruling, and therefore reinstated the assisted-suicide ban, the Michigan Commission began meeting. Within days after its first meeting on July 30, 1883, in which they began to consider whether properly trained physicians were prepared to address treatment and issues of dying patients, Jack Kevorkian assisted in the August 4, 1993, suicide of Thomas Hyde, a 30-year-old ALS patient.

In terms of historical context, this was a perfect storm of three emerging events. One was a suit challenging outlawing assisted suicide. The second was the initial meeting of the Michigan Commission on Death and Dying (which was supposed to advise the legislature as to whether and how to regulate assisted suicide). Third was Jack Kevorkian assisting in a suicide that was at the time unquestionably illegal, by force of a law—a legislative response to his activities. The 1993 indictment (and 1994 Detroit trial) for the assisted suicide of Tom Hyde, under this first temporary assisted-suicide ban of 1993, was thus a response to Kevorkian's activities.[26]

However, Kevorkian and his lawyer, Geoffrey N. Fieger, chose their venue carefully for this test case. Rather than have the first banned assisted suicide take place in Oakland County, with pro-life Chief Prosecuting Attorney Richard Thompson, they chose a county with a pro-choice chief prosecuting attorney, John O'Hair. On August 17, 1993, O'Hair announced at his press conference that he was proposing a piece of (pro-assisted suicide) legislation for consideration by the newly created Michigan Commission on Death and Dying. This was at the press conference he ostensibly called to announce Kevorkian's indictment. O'Hair said that his proposed legislation would allow for assisted suicide under strict, enumerated circumstances for those who were terminally ill and had less than six months to live. Some have asserted that O'Hair was considered by some as the "leader of the pro-assisted suicide group" of the Michigan Commission on Death and

Dying—the same task force created by the same piece of legislation that had first outlawed physician-assisted suicide in 1993.[27] Subsequently, O'Hair took on a prominent role in "Merian's Friends," a group named after Merian Frederick, one of Kevorkian's Oakland patients, for whose assisted suicide Kevorkian was acquitted, after the first 1996 trial; the (unsuccessful) effort of this group to pass a lawful assisted-suicide law.[28]

In O'Hair's press conference on August 17, 1993, he said that "assets that may have been accumulated don't go to the families, surviving spouse, son or daughter . . . they go to the medical profession." Thus, a dying patient's wealth was siphoned off to pay medical bills. In his August 17, 1993, press conference, O'Hair stated, "if Mr. Hyde [the consenting victim] had endured for a few weeks longer, he'd have been force fed, been on a respirator, been comatose." Continuing, he said that what he "found difficult is that he would have to wait that additional two weeks or whatever the time period that may be to go into a comatose state before he could have his voluntary choice fulfilled." In a clear demonstration of his own position, and how it affected his decisions, O'Hair declared:

> I believe that Jack Kevorkian whether he is sincere or not—and I know he is of questioned sincerity—*has tried to bring a focus on this assisted suicide issue and in doing so he has performed a public service*. The disregarding of the law is something I cannot abide, *but at the same time he has performed a public service by his effort and his crusade. I share a concern with Dr. Kevorkian [about] the need to resolve this*. (O'Hair press conference, August 17, 1993; emphasis added)

One fact that both prosecutors noted during August 1993 was that Kevorkian was no longer licensed as a professional physician. (Somewhat surprisingly, Marlinga, chief prosecuting attorney of McComb County where the Hugh Gale case occurred, did not seem concerned by this.) O'Hair took the approach that:

> Kevorkian is not a licensed physician. I'm not sure that the method used [in the assisted suicide of Thomas Hyde] wouldn't be considered cruel by the medical profession. Thomas Hyde over in Belle Isle in a rusted broken down van [referring to the VW van owned by Kevorkian] inhaling carbon monoxide as a means of ending his life—hard to characterize that as dignity . . . [this is the] type of thing we want to avoid. (O'Hair press conference, August 17, 1993)

Moreover, O'Hair advocated methodologies of how to assist in death during an August 19, 1993, interview. These included pills or "lethal injection," a death-penalty methodology that might have been more a reflection of his own profession of criminal law enforcement. As the Kevorkian cases progressed, this was an ongoing theme.

This might be compared to Thompson's view that:

> I think that if Jack Kevorkian left the scene, there would be a more rational discussion by the medical community about the pros and cons of what makes this issue—there are alternatives, such as hospice, [that have] not been discussed. (Thompson interview, August 20, 1993)

O'Hair clearly took a different approach. At his press conference, he stated:

> *Jack Kevorkian has certainly a national, probably an international recognition.* I think I would rather characterize his status as being one of recognition throughout the country, *whether or not he [Kevorkian] is destined for greatness.* (O'Hair press conference: August 17, 1993; emphasis added)

The surprise here was not that Chief Prosecuting Attorney O'Hair acknowledged Kevorkian's fame, but rather that he acknowledged a possibility of the greatness, with regard to the conduct for which he was being indicted. In this case, seeing, hearing and examining the demeanor in person made the difference—the tone O'Hair used was earnest, not ironic or tongue in cheek. In other words, O'Hair viewed Kevorkian as precipitating a social and legal reform movement, while Thompson's view was that Kevorkian was simply acting criminally.

After a trial in Detroit, Kevorkian was acquitted of any wrongdoing in the Hyde case. Perhaps the jurors picked up on the views of the Wayne County prosecutors, thus having a "trickle-down" effect in how the trial was conducted, and on O'Hair's public statements and legislative efforts to the effect that assisted suicide should be lawful. Alternatively, the jury may have been reflecting the emerging social mores of the time and engaged in "jury nullification" (acquittal despite evidence that proves the elements of the crime charged beyond a reasonable doubt). In either event, the jury reached the conclusion that Kevorkian's intent in using carbon monoxide gas was to relieve pain, not to kill; and thus falls within the ambit of the assisted-suicide statutes escape clause for double effect.

As a historical irony, less than one week after Kevorkian was acquitted, the Michigan Court of Appeals, by a vote of 2–1, struck down the state's assisted-suicide ban on procedural ground. However, the Michigan Supreme Court reversed this decision upon appeal on December 13, 1994, and the "temporary assisted suicide ban" stayed in effect.[29] That said, Kevorkian had been in a no-lose situation in Michigan. This is because he had a "friendly" (not antagonistic) prosecutor, and, had he been convicted, would likely not have been subject to a prosecutorial request for prison time. In addition, if the civil and criminal challenges to Section 7 were successful, any conviction would have been null and void.

This all said, Public Act 3, including the assisted-suicide ban, expired on November 25, 1994, further to its sunset clause. The Michigan Commission on Death and Dying concluded its deliberations and issued two reports. One of these was a majority, in favor of lawful assisted suicide, and the other was a minority dissenting report, opposing euthanasia. Wayne County Chief Prosecuting Attorney John O'Hair was a central proponent of the former, while the latter had a coalition including Oakland County Chief Prosecuting Attorney Richard Thompson; during hearings and deliberations, there was marked tension between the two sides.[30] The document itself was considered "A Report by a Group of People," rather than a state document that its title, "Final Report of the Michigan Commission on Death and Dying," sought it to be, because of the sunset of the act.

Some comment on this composition of Michigan Commission on Death and Dying is warranted here, again to place historical context. Its 43 members and alternates included legal and medical professionals (but no judges or legislators), members of the ACLU, AARP, Michigan Hospice Organization, Michigan Hospital Association, Michigan Nurses Association, Michigan Psychiatric Association, National Association of Social Workers–Michigan Division, Right to Life of Michigan, Hemlock (a right-to-die advocacy group) of Michigan, and members of disability rights groups such as the Michigan Association for Retarded Citizens and the Michigan Council on Independent Living. In meetings and deliberations, much of this broke down on pro-choice and pro-life positions. However, unless one was in the room and saw the members, one would not have the historical context that there were no members of AIDS advocacy groups and almost no representation of traditional minorities, who are considered de facto vulnerable populations, particularly when one is mindful of the historical context of the Nazi euthanasia program. The sole member of color was an African American attorney named John Sanford, who was also disabled.[31]

Evan as all this transpired, legislation to reenact a ban on assisted suicide was introduced into the Senate by Sen. Fred Dillingham (SB 1311) and into the House by Reps. Joe Palamara and Ken Sikkema (HB 5968), while Representative Lynn Jondahl introduced a bill to legalize assisted suicide and euthanasia or "aid in dying" (HB 5966). During this period of time, Kevorkian continued to assist in suicide.

By the authority of a companion case to the *Kevorkian/Hobbins*[32] cases in the Michigan Supreme Court, the charges pertaining to Kevorkian's assisted suicides of Marjorie Wantz and Sherry Miller (1991) were reversed and remanded to Judge Breck to try for "open murder." So, too, were charges against Kevorkian reinstated pertaining to the assisted suicides of Merian Frederick and Dr. Ali Khalili, which had been dismissed by Judge Jessica Cooper, when she held that the assisted-suicide law was unconstitutional. Thus, Kevorkian faced trial twice in 1996 in these reinstated cases, regarding two prosecutions each.

The 1996 trials, which took place in March and then in April–May, were for Kevorkian activities in reverse chronological order. That is to say, the first 1996 trial was also tried under the 1993 statutory temporary ban, in a combined trial pertaining to two victims, on different dates. The second 1996 double trial (two victims at the same place and time, but with different methodologies) was for acts that predated the legislatively enacted ban, under a theory of "open murder" or Michigan "common law." Nevertheless, the two cases had the same result—acquittal of all charges. At least in part, this was due to the emerging debate regarding assisted suicide, and also influenced by profound experiences of the participants in the trial. This refers not only to the witnesses at the trials who had personal connections to the decedents, but also to the experiences brought to the trials by the jurors, who are the finders of the facts. Collectively, these acquittals seemed rooted in the concept of "jury nullification"—that is, even though the state adduced sufficient proof of the elements of the crimes charged, and of the causes of death and Kevorkian's intentions, the juries acquitted.

During the March 1996 trial for Kevorkian's assisted suicide of Frederick and Khalili, Bishop Donald Ott of the United Methodist Church, who was the foreman of the jury in the first 1996 assisted-suicide trial, was nearly excluded from the jury pool by Kevorkian's defense attorney Geoffrey Fieger, in this case regarding two distinct allegations. These were Kevorkian's 1993 (post-legislation assisted-suicide ban) aid in the carbon monoxide inhalation suicides of 72-year-old homemaker/ALS patient Merian Frederick and that of 61-year-old Chicago physician/bone cancer patient, Dr. Ali Khalili. Bishop Ott was widely credited as the primary mover of acquitting Kevorkian of assisted suicide in March 1996, and was specifically named in this regard by Judge Jessica Cooper in 1996. Bishop Ott was ultimately a *failed* juror choice on the part of the prosecution. This is because the prosecutors knew of his religious affiliation, but seemingly did not discover that he had written a pro-assisted-suicide article in 1993, believed to be around the time of the death of his family member. In a very real way, this echoed the sentiments of John O'Hair in dealing with Kevorkian's 1994 trial to acquittal for the assisted suicide of Tom Hyde. In addition, Frederick's own minister, Rev. Ken Phifer, essentially offered testimony benefitting Kevorkian. In this, he was joined by Frederick's daughter, who went on to become a speaker and writer in favor of assisted suicide, and to found the group Merian's Friends, which sought to introduce legal assisted suicide in Michigan.

Even had Kevorkian been convicted in March 1996, it is unlikely that Judge Jessica Cooper would have incarcerated him, given that several years later, she allowed Kevorkian to remain at liberty pending bail after a jury convicted him of murder in the second degree of Tom Hyde. During the bail hearing three years later, on March 26, 1999, Judge Cooper and Kevorkian engaged in the following colloquy, which would have been defied credulity in a garden variety (non-euthanasia) murder case:

The Court: Dr. Kevorkian? I understand we have a difference of opinion on this particular subject [of assisted suicide], the law and you. Right?

Dr. Kevorkian: Yes.

The Court: You are pending a sentence by this Court. Can I have your word that there will be no activity during this period of today's date and your sentence?

Dr. Kevorkian: No illegal activity—no unlawful activity. Yes.

The Court: Unlawful in my definition—

Dr. Kevorkian: Yes.

The Court: —not necessarily yours, sir.

Dr. Kevorkian: Well, I mean unlawful in general.

The Court: *No assisted suicide. No injection* [*sic*]. *No anything.*

Dr. Kevorkian: No. No. I kept my word up until now and I'll keep it.

The Court: And you understand the consequences would be severe should there be anything—if you break your bond to me?

Dr. Kevorkian: Yes.

The Court: And that is your word, sir?

Dr. Kevorkian: My word.

The Court: I'll take your word, sir.[33]

However, after the March 1996 trial, Judge Cooper had this to say:

Just as you have zealots on one side, you have zealots on the other. As a judge, I can't comment, take a position, but it's [assisted suicide and euthanasia] highly controversial and they're zealots. *I think [Kevorkian is] a zealot, in every social movement you have zealots, people in your face, nobody likes people in your face and that hurt Dr. Kevorkian . . . [but] then along came change in social relationships.*[34]

During the second 1996 trial, for "open murder" of Marjorie Wantz and Sherry Miller, Kevorkian knew that he had a sympathetic audience, returning to Judge David Breck, who had previously dismissed the charges. Again facing the offices of Richard Thompson, but knowing that the judge was sympathetic, Kevorkian engaged in a variety of antics, including assisting in the suicide of 53-year-old MS patient Austin Bastable on May 6, 1996, on the evening between his direct examination and his cross-examination. Kevorkian did so at the home of Janet Good, who was the president of Hemlock of Michigan, and he administered carbon monoxide.

Again, in an ordinary garden-variety murder trial, engaging in the same conduct while on bail and on trial would ordinarily result in the summary revocation of liberty, yet Judge Breck did not remand Kevorkian. On May 14, 1996, nearly two weeks later, the second jury in nearly as many months acquitted Kevorkian of deliberately causing death. The jury concluded that Kevorkian's intent was to relieve suffering, rather than to kill Wantz and Miller in 1991. During a post-trial press conference, one male juror described the trial period as "five weeks—we have been hostage since April 1." Interestingly, the jury divulged in the press conference that their deliberations were for a total of 12.5 hours (short for a five-week trial) and required only three votes.

The jurors pointedly, repeatedly, and variously stated that what had swayed their opinion. First was the lack of a definition of the "common law" regarding euthanasia and assisted suicide. Second, the cases for which Kevorkian was being tried happened in 1991, with the Supreme Court (of Michigan) issuing a definition for criminal conduct of assisted suicide in 1994; the jury was clearly offended by what it viewed as an *ex post facto* law that had no explanatory backstory. One juror, "Jennifer," enumerated it simply as "1. [T]he prosecutor didn't make an attempt to explain the common law[,] and 2. [H]ow should a decision in 1994 affect Dr. Kevorkian in 1991?"[35]

Thus, the jury did not seem to accept that while they were the judges of the facts, the judge was required to provide instructions regarding the law, but not required to provide a "written law" that was in effect when the Wantz/Miller assisted suicides took place in 1991. What the jurors apparently wanted was black-letter law, literally in the printed form in front of them. The law was not provided because the case was tried under a theory of common law of homicide (as upheld and required by the combined cases of *People v. Kevorkian* and *Hobbins v. State*; that was the case in which the Michigan Supreme Court reversed and remanded an earlier dismissal by Judge Breck for Kevorkian to be tried), as predating the first statutory ban on assisted suicide in Michigan. The statutory ban was passed in 1993, and the "Breck" jury perceived the case as an assisted-suicide "common law" passed in 1993 reaching back to 1991. Indeed, it also seemed that the jury imposed a burden on the prosecution to provide and explain the law, while it would have been reversible prosecutorial and fair trial error to usurp the judge's province, had the prosecutor so done. Because, as the jurors stated in their press conference, "of course" they knew about Dr. Kevorkian before his trial, and it would appear that the jury, in effect, deputized him as an expert witness, as well as a defendant. Since the jury felt Kevorkian did actually make an affirmative showing during his testimony in the second 1996 case, they acquitted him.[36] This stood in stark contrast to the 1999 Kevorkian trial, in which he acted as his own counsel, and opened and closed the case, but did not testify.

The other factor, while not identified by the jury until the end of their press conference, was that what really touched them and impacted their decision-making process was the testimony of the families of the decedents who had been served by Dr. Kevorkian—Sherry Miller's mother and her best friend, and Marjorie Wantz's husband. This, along with the videotapes of the women, the jurors found compelling. Indeed, they were so compelling that during deliberations, one male juror, "Vince," asked Cameron Beedle to "please turn around those pictures of those dead women."[37]

Thus, for the acquitting jury in the second Kevorkian trial in 1996, the factors that were most prominent were first, a lack of clarity on a non-statutory "common" law. Second was the appearance of a judicial or prosecutorial vendetta by criminalization of the conduct after the fact. Third was the impact of the families and friends of the women who had died, despite their non-terminal illnesses. However, this acquittal occurred less than two months after the acquittal in the Frederick/Khalili assisted suicides (by carbon monoxide inhalation) that took place in 1993, when the temporary statutory ban (permanent by the time of these trials) was in place.

So it was that the prosecutor's office was faced with two separate jury verdicts acquitting Kevorkian, on two separate legal theories of the cases, in two separately conducted trials. First was the acquittal of statutory assisted suicide (Judge Copper's 1996 Kevorkian trial, regarding two statutory 1993 cases under the first assisted-suicide ban, based on a successful defense theory of double effect of alleviating suffering by use of the carbon monoxide inhalation). Second was the acquittal under a theory of common-law murder (Judge Breck's Kevorkian trial, regarding two 1991 common-law murder cases that occurred, were first prosecuted prior to the statutory scheme).[38]

This all said, even had Dr. Kevorkian (who still had a license to practice medicine when the Wantz/Miller cases took place, although the cases were the basis for his loss of licensure) been convicted, he would not have been facing a prison sentence for these acts of euthanasia. Judge Breck, on May 15, 1996 (48 hours before my interview with Judge Cooper), also commented on zealotry, but extended it to sentencing issues. Judge Breck made seemingly contradictory comments in rapid succession—that Kevorkian was "tough to handle" and that it:

> would have been tough to fashion a sentence, I'd give a year of probation, suspended, no penalty whatsoever. . . . but his disdain for the courts, made me wonder a little bit if I'd follow through.[39]

The sentence Judge Breck contemplated for the "tough to handle" Kevorkian was more in accord with misdemeanor shoplifting than open murder. Thus, Judge Breck would have been basing his potential sentencing structure for Kevorkian upon how he viewed the right-to-die debate or assisted-suicide movement, rather than upon the defendant. This view of Kevorkian regarded the issue, not the charges before him or the defendant's conduct, and certainly not any sentencing guidelines. An influence on Judge Breck's views may have been his wife's death some years earlier, after a lengthy illness and her experiences with hospice, of which Judge Breck was a great proponent.

1997–1999: THE KEVORKIAN INJECTION TRIALS AND THE *60 MINUTES* INTERVIEW

In 1997, Kevorkian was briefly on trial for the 1996 assisted suicide of Loretta Peabody, in rural Ionia County. Janet Good, president of Hemlock, was indicted as a coconspirator; however, because Good had terminal pancreatic cancer (and indeed died in August 1997, in an assisted death by lethal injection, with Kevorkian's assistance) the Ionia County prosecutor, Ray Voet, dropped the charges against her. Peabody's death certificate read that "the immediate cause of death" was "Death by IV injection," and also listed as a "significant condition contributing to death," but not resulting in the underlying cause of death given was multiple sclerosis (MS).[40] The death certificate provided a strong prosecutorial offense—that while assisted suicide was the charge, there is no statute of limitations on murder. However, there was an equally strong defense—of an intervening cause of death by the underlying illness, an opportunity that would, for all intents and purposes, provide the successful defense of Hermann Sander in the 1950s, that his patient was "already dead."

Voet noted that he had, prior to trial, brought a motion for "decorum," i.e., for rules of conduct and propriety in court at trial. An unintended consequence was that instead Voet's motion might have been an inadvertent road map for Fieger, whose opening statement was so inflammatory as to provoke the unusual measure of a mistrial. Voet ruefully commented that he had observed the proceedings even as he participated in them during his and Fieger's opening statements. His inner dialogue was, "here it comes, judge;

there it goes; see, judge, it just went by"[41] when referring to Fieger's lack of decorum and misconduct during his opening to the jury; Voet likened this to "jury arson" of an all-white, seemingly conservative jury, in which Fieger found no friendly faces for either his client or his issue. Kevorkian's lawyer would, however, have found a friend in the judge, who stated that he did not know if he would have imposed a jail sentence on Kevorkian, in light of Kevorkian having had "no prior convictions" on his record (notwithstanding the dozens of assisted-suicide participations).[42]

While Kevorkian was awaiting trial with regard to his participation in Peabody's death, a new chief prosecuting attorney, David Gorcyca, dismissed all remaining charges against him. Thus, he summarily dismissed all the indictments that outgoing prosecutor Richard Thompson had filed in the wake of a lost election—an election lost partly because Gorcyca had campaigned on a platform that prosecuting Kevorkian for his assisted death activities was an exercise in futility. On January 6, 1997, Gorcyca told the press:

> Hypothetically, we can charge and convict Dr. Kevorkian 100 times, but since we aren't making law nor are we setting precedents, what are we really accomplishing other than wasting a lot of taxpayer money?[43]

Later in 1997, a group called Merian's Friends (in honor of Merian Frederick, of whose death Kevorkian was acquitted of assisting in, during the March 1996 trial) sought to have Proposal B, allowing for lawful assisted suicide by prescription, placed on the ballot. The "Terminally Ill Patient's Right to End Unbearable Pain or Suffering" ballot initiative provided for lethal prescription, by definition requiring a licensed physician, and thus excluding Kevorkian. The proposal also provided for protections similar to those passed by Oregon voters with the "Death with Dignity Act" in 1994. These included repeated witnessed voluntary requests in writing and/or on video, the right to withdraw the request, at least two doctors certifying terminal illness with less than a six-month life expectancy, confidentiality provisions for patients and against the public and the press, a waiting period, and that the patient be the person to self-administer the lethal dose. The proposal ultimately failed in November 1998, by a margin of 71 percent to 29 percent, facing what Merian's Friends called opposition by the Catholic Conference and Michigan Right to Life.

Although Kevorkian would have been excluded from practicing assisted suicide under the protocols of Proposal B, its defeat propelled him to escalate further. After participating in over 130 hastened deaths, Kevorkian, through a media intermediary, arranged for a tape of him administering euthanasia to be delivered to Mike Wallace of the CBS newsmagazine *60 Minutes*. The tape showed Kevorkian administering euthanasia to a 52-year-old end-stage

ALS patient named Tom Youk. The date of death was September 17, 1998, but the tape was not sent until after Proposal B was defeated on Election Day. On November 22, 1998, CBS aired a segment entitled "Death by Doctor," in which Kevorkian narrated the tape and was also interviewed by Wallace. The network broadcast an interview, as well as a consent colloquy, of Youk, conducted by Kevorkian. Then the 18-minute-long segment showed Kevorkian administer drugs by intravenous injection, while he narrated Youk's last moments. CBS also interviewed Youk's family, who attested to his pain and suffering, as well as to his desire for assisted suicide; Kevorkian and Youk ultimately agreed to euthanasia by injection, both knowing that the media and the general public would almost certainly see the tape. During the *60 Minutes* interview, Kevorkian told Wallace:

> I've got to force them (the prosecutors) to act. They must charge me. Because if they do not, that means they don't think it's a crime. Because they don't need any more evidence do they? Do you have to dust for fingerprints on this[?][44]

Three days later, Kevorkian was charged with three felonies—murder in the first degree, assisting in a suicide, and delivery of a controlled substance. Less than four months later, he was on trial before Judge Cooper again. Trial prosecutor John Skrzynski, who had faced Kevorkian in March 1996 (also before Judge Cooper), had the 18-minute-long "Death by Doctor" tape pruned to 9 minutes, and successfully excluded any evidence of Youk's pain and suffering from the trial. This, in effect, meant excluding Youk's widow, Melody, and brother, Terry, from testifying on behalf of Kevorkian.

One aspect of the 1999 Kevorkian trial was that there were competing contentions as to the cause of the death of Tom Youk in September 1998. At first blush, this may sound absurd, given the nine-minute version of the prosecutor's clips in evidence, of Kevorkian's tape and the *60 Minutes* program interview with Mike Wallace. During this, Kevorkian narrated and provided commentary (in both the original program and the prosecutor's clips) as he injected Youk with a lethal cocktail, strikingly similar to the protocol for death penalty by lethal injection. However, there was a possible (albeit seemingly far-fetched) alternative conclusion the jury could have reached—that Youk died from the underlying disease of ALS (described earlier in the chapter) in the course of the events taped, and was "already dead," similarly to New Hampshire's Dr. Hermann Sander in the 1950s, and Winchester, England's Dr. Nigel Cox in 1992. If the jury concluded that Youk had died of the underlying ALS, that would have gone beyond the standard of reasonable doubt for an acquittal to an actual doubt; however, had the jury embraced the possibility of such a death, it would have been legally required to acquit Kevorkian, due to a lack of proof of the element of cause of death.

Oakland County Medical Examiner Ljubisa L. Dragovic testified at length about the September 17, 1998, autopsy (which he witnessed) and his viewing of the Kevorkian tape. In short, he testified that there was enough secobarbital (a barbiturate to induce sleep) present in Tom Youk's body to kill him within a few hours and further that the paralysing muscle relaxant anedctadine was present in a sufficient quantity to kill him within five to eight minutes. However, the autopsy could not determine whether the potassium chloride had been injected because that is present in the body after red blood cells die.[45] Kevorkian commented on Youk's death for Mike Wallace in a portion of the prosecutor's clips of the *60 Minutes* program shown on March 23, 1999, to, and transcribed for, the jury (as shown below, as verbatim and as styled from the trial transcript, with indents and double indents):

K[evorkian]: And we're ready to inject. We're gonna inject you in your right arm now. Okay? Okie-dokie.

K: Sleepy Tom? Tom are you asleep? Tom are you asleep? You asleep? He's asleep.

. . .

W[allace]: And this—

K: Paralyses the muscles.

W: But he's still alive?

K: He's still alive, but, ah, and that's why I . . .

W: Now I can see his breathing just a (inaud).

K: That's why I have to, you know, now that there lack of oxygen getting to him now, but he's unconscious deeply so it doesn't matter.

W: Is he dead now?

K: I don't, he's dying now, cause his oxygen's cut off, he can't breathe. So now I'll quickly inject the potassium chloride to stop the heart.

K: Now there's a straight line [*sic*].

W: He's dead.

K: Yep. The heart has stopped.

K: Straight line. The cardiogram will be turned off.[46]

Taken together, the testimony of Dragovic plus the narration by Kevorkian could be (and presumably was) construed by the jury as sufficient evidence to prove the element of causation in Tom Youk's death beyond a reasonable doubt. The jury convicted Kevorkian of second-degree murder and delivery of a controlled substance, which was upheld on appeal in the final *People v. Kevorkian*.[47] Perhaps it is an irony that the intellectual aspect of cause of death may have been well fought between the two doctors, but understood by few in the courtroom. In addition, the question was effectively forfeited when Kevorkian, acting as his own lawyer in this final trial, argued only the question of whether he had intent to commit murder.[48]

The jury, having repeatedly seen, heard, and read the nine-minute version of the "Death by Doctor" segment, determined that Kevorkian intended to commit murder, rather than to relieve pain and suffering. Unlike previous cases, the jury and jurors declined to hold a press conference or to give interviews, leaving the public to speculate about the reasoning that led to the verdict of guilty. Judge Cooper initially allowed Kevorkian to be at liberty pending sentence (which many took as an indication that she would not sentence Kevorkian to any time in prison); Kevorkian then told probation authorities that they could not stop him. Judge Cooper reacted strongly to this:

> No one is unmindful of the controversy and emotion that exists over end of life issues and pain control. And I assume that the debate will continue in a calm and reasoned forum long after this trial and your activities have faded from public memory. But this trial is not about that controversy. The trial is about you, sir. It was about you and the legal system, and you have ignored and challenged the Legislature and the Supreme Court. . . .
>
> Now another consideration, and perhaps a stronger factor in sentencing, is deterrence. This trial was not about the political or moral correctness of euthanasia, it was all about you, sir. It was about lawlessness. It was about disrespect for a society that exists and flourishes because of the strength of its legal system. No one, sir, is above the law—no one.
>
> So let's talk just a little bit more about you specifically. You were on bond to another judge when you committed this offense. You were not licensed to practice medicine when you committed this offense, and you haven't been licensed for eight years. And you have the audacity to go on national television, show the world what you did, and dared the legal system to stop you. Well, sir, consider yourself stopped.[49]

Kevorkian's assisted-suicide and euthanasia career ended with an incarceratory sentence of 10–25 years (for second degree murder) and a concurrent sentence of 7 years (for delivery of controlled substances).[50] In this chapter, the prosecutions also provided cases in which to place nearly every historically raised social, medical, and legal question of the 1900s, a microcosm of a century. One study suggested that perhaps it was not that Kevorkian was prolific, so much as that he was repeatedly and increasingly public and self-publicizing, in his activities.[51] Surveys of Michigan physicians' attitudes and conduct lend support to this argument.

CONCLUSION

In any event, from the early interviews with Chief Prosecuting Attorneys Thompson and O'Hair, representing the pro-life and pro-choice views, until the final sentence, a robust debate developed in Michigan. This included consideration of lawful and unlawful assisted suicide, the stress between family pressures that might stress vulnerable populations, and what Thompson, predicting the debates of the post-millennial society, noted was "an unspoken and unwritten message going out to every elderly person in our society that once you become about 80 and start to become infirm, don't waste our money—kill yourself, because your life is not gonna be real." In the Michigan cases, some family members and clergy went from being reserved to being activists. Media precipitated, as well as reported, on the Kevorkian cases.

Jack Kevorkian himself went from being viewed (largely) as a compassionate doctor willing to place himself at risk for the benefit of his patients, short term though they were, to being a folk devil who was pilloried by the criminal justice system. In no small measure, he went from being viewed in the beginning of the decade like England's Dr. Nigel Cox (that is, as a sympathetic doctor responding to patient needs, and placing himself at legal risk) to being viewed as more akin to Britain's Dr. Harold Shipman (that is, a prolific serial killer who found easy victims, willing in one way or another, even if Shipman's patients intended to consent to treatment, rather than to death). At the end of the decade, the debate ranged across pro-choice and pro-life, across urban jurisdictions and rural, across civil litigation and criminal prosecution, and informed (as well as was informed by) other jurisdictions, as Chapter 6 will develop.

NOTES

1. *Roe v. Wade*, 410 U.S. 113 (1973).
2. Herring, Mark Y., *Historical Guides to Controversial Issues in America: The Pro-Life/Choice Debate* (Westport, CT: Greenwood Press, 2005), 83.
3. Mazer, Adam, *You Don't Know Jack*, directed by Barry Levinson (HBO Films, 2010).
4. The movie was based in part upon Neal Nicol and Harry Wylie, *Between Dying and the Dead: Dr. Jack Kevorkian's Life and the Battle to Legalize Euthanasia* (Madison: University of Wisconsin Press, 2006). Nicol was a friend who was Kevorkian's assistant for some of the assisted suicides, and Wylie is a friend of Kevorkian's.
5. Pappas, Demetra M., "The Politics of Euthanasia and Assisted Suicide: A Case Study of the Emerging Criminal Law and the Criminal Trials of Jack 'Dr. Death' Kevorkian" (PhD. diss., London School of Economics and Political Science, 2009).
6. *People v. Kevorkian*, 205 Mich.App. 180 (Court of Appeals of Michigan, May 10, 1994), reversing, Case Number 92-115190-FC, in which Judge David Breck dismissed murder charges brought against Kevorkian, and remanding the case for trial.

7. *Id.*, 183.
8. *Id.*
9. *Id.*, 191.
10. 211 Mich. 187 (1920).
11. 211 Mich. 192.
12. 211 Mich. 194–95.
13. *People v. Kevorkian*, Case No-CR92-115190-FC and 92 –DA-5303-AR, Opinion and Order (Cir. Ct. Oakland County, July 21, 1992) (Breck, J., writing), p. 12 (unreported decision).
14. *Id.*, 13.
15. *Id.*, 10. While the higher courts of Michigan would determine that the case was otherwise, they did so on the basis of the legislative enactment that followed after Judge Breck dismissed these cases.
16. *Id.*, 16.
17. *People v. Campbell*, 124 Mich.App. 333 (Mich. Ct. App. 1983).
18. *Hobbins v. Attorney General*, No. 93-306-178 CZ (Cir. Ct., Wayne County, May 20, 1993) (Stephens, J., writing).
19. *People v. Kevorkian*, Case Number 93-129832-FH and Case Number 94-130248-FH (Cir. Ct. Oakland Cty., January 27, 1994) (Cooper, J., writing). In this decision, Judge Cooper dismissed charges against Kevorkian emanating from his involvement in the deaths of Merian Frederick and Dr. Ali Khalili; these charges were reinstated on appeal, and tried to acquittal in 1996.
20. *Id.*, p. 14.
21. *Hobbins v. Attorney General*, No. 93-306-178 CZ (Cir. Ct. Wayne Cty., May 20, 1993) (Stephens, J., writing), p. 2.
22. MCL 752.1027(a), (b).
23. *Hobbins v. Attorney General, op. cit.*, May 20, 1994.
24. Volume II (1994), p. 18.
25. *Hobbins*, p. 10 (Stephens, J., writing). This case was subsequently reversed, but it represents a novel issue and decision of the time.
26. According to the interview statements of Senator Fred Dillingham, R-Fowlerville (August 23, 1993), among others, of "Section 7," the 18-month temporary ban outlawing assisted suicide was commonly referred to as the "anti-Kevorkian law." Pappas, "The Politics of Euthanasia and Assisted Suicide."
27. Zalman, Marvin, John Strate, Dennis Hunter, and James Sellars, "The Michigan Assisted Suicide Three Ring Circus: The Intersection of the Common Law and Politics," *Ohio Northern University Law Review* 23, no. 3 (1997): 924.
28. Proposal B was defeated in November 1998, only days before Jack Kevorkian arranged to have someone deliver a tape of the September 1998 euthanasia of Tom Youk to CBS for broadcast on *60 Minutes*, precipitating the final trial to its 1999 conviction.
29. *People v. Kevorkian*, 447 Mich. 436 (1994).
30. Personal observation, March 4, 1994. In addition, there were demonstrations at the hearings, as marked on March 4, 1994, where the interest group Not Dead Yet

staged a demonstration, organized in part by Senator Dillingham and Ed Rivet, legislative director of Right to Life of Michigan.

31. There was an interesting introduction to my interview of John Sander in March 1994. The first thing he asked me was whether I was interviewing him because he was black or because he was disabled. Taking an approach equally honest and practical, I asked him which response would result in him granting the interview. While my reply elicited a laugh from each of us, he had obviously already had in mind to give me the interview, as I had been invited to his office for that purpose. However, his point was well-taken—minorities were largely socially excluded from the process, other than Sanford. Sanford, who worked on disability issues, expressed concern over the lack of African American, Hispanic, and minority voices and perspectives repeatedly. While that is a footnote in this chapter, it is nonetheless noteworthy as an example of the exclusion of minority populations.

32. *People v. Kevorkian*, 447 Mich. 436 (1994).

33. Kevorkian transcript, March 25, 1999, vol. IV, pp. 11–12; emphasis added.

34. Cooper interview, May 17, 1996; emphasis added.

35. Breck Jury press conference, May 14, 1996.

36. *Id.*

37. Beedle interview, May 19, 1996.

38. These seemed to have been lessons learned by the prosecutor's offices by 1999. This argument is supported by the fact that Kevorkian was charged with both the assisted-suicide and euthanasia murder of Tom Youk only after a new law permanently banning assisted suicide was passed, and the vigorous and successful opposition to Kevorkian being allowed to call Tom Youk's wife Melody and brother Terry to testify to Youk's pain and suffering.

39. Breck interview, May 15, 1996.

40. Certificate of Death of Loretta Peabody 1387861, dated September 3, 1996.

41. Voet interview, June 13, 1997.

42. Miel interview, June 13, 1997.

43. Gorcyca press conference, January 6, 1997.

44. Kevorkian trial transcript, March 23, 1999, *60 Minutes*, at 30.

45. *People v. Kevorkian*, 248 Mich.App. 373 (2001).

46. Kevorkian trial transcript, March 23, 1999, 28–29.

47. *People v. Kevorkian*, 248 Mich.App. 373 (2001).

48. The charge of assisting in a suicide was dropped by the prosecutor pre-trial, so that Kevorkian was tried only for the higher count of murder, along with drug delivery charges.

49. Kevorkian sentencing, April 13, 1999: 37–38.

50. *People v. Kevorkian*, 248 Mich.App. 373 (2001).

51. One such example of this was an article by Jerald G. Bachman et al., "Attitudes of Michigan Physicians and the Public toward Legalizing Physician-Assisted Suicide and Voluntary Euthanasia," *New England Journal of Medicine* 34 (1996): 303–9.

6

The 1990s: The Non-Michigan Parallel Text of Doctor Prosecution and Initial Failed Legislative Efforts

INTRODUCTION

In 1990, before Jack Kevorkian became the face of assisted suicide, the U.S. Supreme Court considered the case of a young woman named Nancy Cruzan. Cruzan was in a 1983 automobile accident that left her in a persistent vegetative state, from which she would never awaken. By a decision of 5–4, the court in *Cruzan v. Director, Missouri Department of Health*[1] ruled in June 1990 that where there was clear and convincing evidence that a person in a persistent vegetative state would not have wanted to have been maintained in that condition, life support could be removed. In other words, the patient would be allowed to die a natural death, rather than be maintained by medical technology, if there was clear and convincing evidence the now-incompetent patient would have wanted the withdrawal of life-sustaining medical treatment.[2] Although this was in fact a case about the right to refuse medical treatment, it became widely known as a "right-to-die" case. In a sense, this opened the door to kitchen table conversations across the United States regarding how people wanted to be treated (or not) if there was a catastrophic event that left them in a persistent vegetative state. It also opened a second door for many to have a conversation about choices to make while still competent, but facing terminal illness. Last, a third door was opened—for patients to begin to have conversations with family and with medical practitioners about taking control of the time and manner of their deaths, if they were terminally ill or if they had a chronic, degenerative illness (such as MS,

ALS, Huntington's disease, or AIDS, an illness that had not existed until less than a decade earlier).

In the wake of Jack "Dr. Death" Kevorkian's 2011 death, members of the media (including Barbara Walters, who interviewed him a number of times) eulogized him as being responsible for assisted suicide becoming legal in three states—Oregon, Washington, and Montana.[3] However, in none of these was Kevorkian cited positively or consulted for his input. As discussed in the previous chapter, Kevorkian's own state, Michigan, created a criminal law prohibiting assisted suicide.

As an initial matter, the path to Oregon's 1994 legislation, its 1997 implementation and its subsequent post-millennium survival of a 2005 U.S. Supreme Court challenge, will be discussed in the next chapter, as spanning its own decade of change for the millennium. However, in Washington State, some concluded that Kevorkian's activities in the October 23, 1991, Wantz/Miller cases (of which Kevorkian was acquitted, in 1996, and where the decedents suffered from a then-vaguely worded and unspecified pelvic disorder and multiple sclerosis, respectively) barely a week before the November 5, 1991 election, he was a factor in the failure of Washington's assisted-suicide voter Initiative 119, because "Jack Kevorkian ha[d] given a personal face to the philosophical and theological arguments against mercy killing and assisted suicide [in contrast to] the somewhat abstract contention that Initiative 119 would open the door to great abuses and myriad slippery slopes."[4] Moreover, Washington State doctors and patients unsuccessfully sought to have physician-assisted suicide declared legal by the federal judiciary, in a case that went before the U.S. Supreme Court, *Washington v. Glucksberg*.[5] The companion case, emanating from New York, of *Vacco v. Quill*,[6] was also unsuccessful in the U.S. Supreme Court, although it was prompted in part by the failed prosecution of hospice physician Timothy Quill.

Montana did not further legal change for over a decade after Kevorkian's activities were, to use Judge Jessica Cooper's word, "stopped."[7] *Baxter v. State*,[8] the Montana Supreme Court ruling that provided, by a narrow majority of 4–3, for physician-assisted suicide as a matter of state constitutional law, was decided December 31, 2009, a decade after Kevorkian had been convicted of, and served prison and been paroled for, the 1998 murder of Tom Youk.

In other words, while Kevorkian may have dominated local, national, and even international news, as the firebrand of the assisted-suicide and euthanasia debate, it was not to the exclusion of other jurisdictions. And, while he may have brought attention to the debate regarding assisted suicide and been the American poster boy for euthanasia, and placed in historical context, it is inaccurate to say that he was the reason that assisted suicide became legal in three states.

This said, the 1990s was a decade of change, from coast to coast, from local elections to U.S. Supreme Court contemplation. Even where the result of debate was to leave the official status quo unchanged, the mere fact of open discussion of matters previously private (within families) and conducted within the hidden shadows of the law (by doctors) showed a sea change—taking end-of-life matters from the private into the public spheres. Some of these were initiated prior to Kevorkian, but almost all either informed the Kevorkian cases or vice versa, with the exception of a very few events that took place before Kevorkian made euthanasia and assisted suicide a regular front-page news event.

DOCTOR PROSECUTIONS TO COMMENCE A DECADE OF EFFORTS FOR LEGISLATION AND LITIGATION

As with the 1950s, there were doctor prosecutions that precipitated the discourse for change. There was, of course, Kevorkian activities, starting with the 1990 assisted suicide of Janet Adkins, and the legislation and litigation that followed in Michigan, discussed in the previous chapter. However, there was also the 1991 New York prosecution of respected hospice physician Timothy Quill, who gave a leukemia patient, "Diane," a prescription for a barbiturate overdose, which she took to commit suicide. Nearly contemporaneously, there was the prosecution of rheumatologist Nigel Cox in England's Winchester Crown Court for attempted murder of an imminently terminally ill patient (whose body was unavailable for autopsy, as she had been cremated). Rather than simply precipitate a hoped-for social movement and theoretical writings, as did the doctor prosecutions in the 1950s, both of these furthered national (indeed, international) debate, legislative contemplation, and legal exploration. Unlike the 1950s, however, the 1990s saw doctors on both sides of the Atlantic Ocean bringing suit for declaratory judgments to the effect that they would not be criminally prosecuted, or held civilly liable, if they engaged in conduct that would shorten the lives of those in persistent vegetative states and those suffering from terminal illness. These harkened back to the cases of the early right-to-die litigation of the 1970s, but with a much heavier emphasis upon the patient's right to self-determination, and the physician's duty to honor patient autonomy.

In addition, these doctors were dealing with a more sophisticated client base of patients, which suggested the sort of patient activism that would follow during the 1990s. Some terminally ill patients, such as Timothy Quill's patient "Diane," took control of their lives and their deaths—or dying processes. Quill, an internist, specializes in hospice (palliation or relieving suffering precipitated by the physical, emotional, spiritual, and social symptoms of the terminally ill). In March 1991, an essay Quill wrote, entitled "Death and Dignity: A Case of Individualized Decision Making," appeared in the

New England Journal of Medicine. The article described the doctor-patient relationship of Quill and a patient "Diane," who was diagnosed with myelocytic leukemia, an aggressive illness which then had a 75 percent fatality rate. Quill wrote of Diane's difficult life battles in which she had overcome alcoholism and depression, managed to forge stronger family relationships with her husband and adult son, and built a business and artistic practice. In other words, Quill painted a portrait of a patient who wanted to live and had much to live for, at the time of her diagnosis.

Diane repeatedly and enduringly made the choice not to undergo chemotherapy, and persistently declared that she wanted to spend the remainder of her life with her husband and son. While Quill was surprised and dismayed by this decision, he wrote that it was not uninformed—indeed, Diane had consulted with a variety of hematologists, oncologists, and a psychologist, and was adamant in this position. Quill wrote that "[j]ust as I was adjusting to her decision, she opened up another area that would stretch me profoundly. It was extraordinarily important to Diane to maintain control of herself and her own dignity during the time remaining to her."[9] In short, Quill wrote, "[w]hen this was no longer possible, she wanted to die."[10] Quill, referring to his role as a former hospice director, attested that he knew "how to use pain medications to keep patients comfortable and lessen suffering [and that he] explained the philosophy of comfort care, which [he] strongly believe[ed] in."[11]

Diane's response was proactive and demonstrated a more participatory response to medical decision making that began to rise during the 1990s, especially among populations where there was terminal illness (such as cancer) or illness of chronic debilitating illness, illness that would result in death barring an intervening cause of death, such as an accident. Diane demonstrated a "preoccupation with her fear of a lingering death [that] would interfere with Diane's getting the most out of the time she had left until she found a safe way to ensure her death."[12] Quill wrote that he "feared the effects of a violent death on her family, the consequences of an ineffective suicide that would leave her lingering in precisely the state she dreaded so much."[13] Additionally, Quill wrote that he was concerned about "the possibility that a family member would be forced to assist her, with all the legal and personal repercussions that would follow."[14]

Quill recommended that Diane seek information available from the Hemlock Society, and that a week after this, Diane "phoned me with a request for barbiturates for sleep."[15] Because this was what Quill called an "essential ingredient in a Hemlock Society suicide,"[16] he asked Diane to visit his office for a consultation. Quill wrote that "[s]he was more than willing to protect [Quill] by participating in a superficial conversation about her insomnia, but it was important to [Quill] to know how she planned to use the drugs and to be sure that she was not in despair or overwhelmed in a way that might color her judgment."[17] This consultation predicted the sort of regulatory protection that would be legislated in Oregon later in the decade, when Oregonians voted by 51.3 percent to 48.7 percent in 1994 to enact Measure 16,[18] which became the Oregon Death with Dignity Act in 1997. The way in which Quill analyzed the meeting was as follows: "it was apparent that she was having trouble sleeping, but it was also evident that the security of having enough barbiturates available to commit suicide when and if the time came would leave her secure enough to live fully and concentrate on her present."[19]

Placed in historical perspective, the mere discussion Quill described in this essay demonstrated a collaborative decision-making process between doctor and patient, a paradigmatic shift from the paternalistic model in which physicians made decisions for the patient, and patients had the option of giving or withholding consent. Diane in fact received the prescription from Quill, who wrote it "with an uneasy feeling about the boundaries [he] was exploring—spiritual, legal, professional, and personal,"[20] but he also felt that he was "setting her free to get the most out of the time she had left, and to maintain dignity and control on her own terms until her death."[21]

Quill's contention was that by simply having possession of the lethal prescription, Diane was able to spend several months with family and friends relatively free of anxiety about her death, although she battled severe exhaustion and bone weakness. She did, as she promised, meet with Quill for a final, tearful consultation prior to using the drugs to end her life. When her husband called Quill to tell him that Diane had died, her husband told how she had been able to say goodbye to their son and to him before she took the barbiturates. Nearly 15 years later, there would be empirical research findings in the Netherlands, albeit in euthanasia cases rather than assisted suicide, that being able to say goodbye quantifiably mitigated the experience of loss. Swarte, van der Lee, van der Bom, van den Bout, and Heintz produced a quantitative study published in 2003, but one which had as the start of its longitudinal data gathering the year 1992—the year after Diane's death, assuming that Quill's 1991 essay was written contemporaneously with the time of occurrence. The Dutch study was of 189 bereaved families and close friends of terminally ill cancer patients who died by euthanasia versus 316 bereaved family members and friends of comparable cancer patients who died a "natural death" between 1992 and 1999 in the Netherlands.[22] The Dutch researchers concluded that, among other things, "grief experienced by family members in suicide cases differs from grief after euthanasia, mainly because relatives have had the opportunity to say goodbye, which is seldom the case in suicides [and the study authors posited that] physician assisted suicide should be expected to resemble euthanasia on this point, because it will also usually be announced."[23] Indeed, Quill's own writing suggests that Diane's family had well and long prepared for her loss, and that saying goodbye had been important—important enough to report to the doctor, although she was alone at the final moment (with the suggestion that this was so as to exempt her family from criminal liability).

The Dutch give retrospective context to Quill's writing about the experience of Diane's family. As for Quill himself, he wrote that "Diane taught me about the range of help I can provide if I know people well and if I allow them to say what they really want. She taught me about life, death and honesty and about taking charge."[24] Quill also wrote that while the field of medicine has "measures to help control pain and lessen suffering, to think that people do not suffer in the process of dying is an illusion."[25] He asked "whether Diane struggled in that last hour, and whether the Hemlock Society's way of death by suicide is the most benign."[26]

This all said, Quill's participation in the life and death of Diane did not end at her death. Indeed, it was the next paragraph of the Quill essay that perhaps furthered the assisted-suicide debate the most. Quill himself "called the medical examiner to inform him that a hospice patient had died. When asked about the cause of death [he] said, 'acute leukaemia.' [The medical examiner] said that was fine and that we should call a funeral director."[27] This statement was neither accurate nor complete, as Quill went on to note, "[a]lthough acute leukaemia was the truth, it was not the whole story."[28] Quill went on to justify this to any readers of the essay, as well as the readership of the medical journal as a group, "[y]et any mention of suicide would have given rise to a police investigation and probably brought the arrival of an ambulance crew for resuscitation. Diane would have become a 'coroner's case.' "[29] Placed in context, what Quill meant was that the matter would, rather than remain a private matter, been exposed to public scrutiny. However, the publication of the essay invited public scrutiny by medical professionals, and by the criminal justice system.

In this regard, Quill wrote that "[t]he family or [Quill] could have been subject to criminal prosecution, and [Quill] to professional review, for [their roles] in support of Diane's choices."[30]

In the next sentence of Quill's essay, he authored what might most appropriately be considered the opening salvo of the assisted-suicide debate of the 1990s. In particular, Quill wrote that "although [he] truly believe[d] that the family and [Quill] gave best care possible, allowing her to define her limits and directions as much as possible, [Quill was] not sure that the law, society or the medical profession would agree."[31] This quote of this first article of represented the first openly acknowledged and reported act of clear physician-assisted suicide.

Indeed, when Diane's real name, 45-year-old Patricia Diane Trumbull, and the date of her May 19, 1990, death came to light,[32] the historical record would show that Quill's prescription-assisted suicide of Diane occurred prior to Jack Kevorkian's first acknowledged hastened death—that of Janet Adkins, by intravenous injection, on June 4, 1990. The record is clear that Diane was the final actor in her death, choosing when and how to take the barbiturates, on her own terms and in her own time. This is in some contrast to Kevorkian's activity in the death of Janet Adkins, in which he was present and also may, some believe, have been the final actor to administer the drugs by use of a suicide machine, in what would then have been a case of euthanasia. Quill consistently sought to distinguish himself from Kevorkian, and hoped that the debate following his case (which would prove to be his sole case of assisted suicide) would be viewed in contrast to that that of pathologist Kevorkian (who would go on to participate in over 130 cases of assisted suicide and euthanasia).[33]

There are a few immediate contrasts that contextualize the historical roles of these two cases—Diane had been a long-time patient of Quill's, while Adkins knew Kevorkian for only a few days at the time of her death. Quill had counseled Diane repeatedly and enduringly to consider alternative treatments that might forestall, if not cure, her extremely aggressive and life-threatening cancer; whereas Janet Adkins, who was showing early symptoms of Alzheimer's disease, was not imminently terminally ill and had not slipped into the middle or late stages of Alzheimer's, which are most frequently characterized by loss of lucidity, among other symptoms. While both Diane and Janet Adkins were middle-aged white women, as time went on, Jack Kevorkian would assist in the deaths of eight more middle-aged white women before assisting in the suicide of any men (and, ironically, would be tried for his role in the deaths of three men and four women, although some 72 percent of his patients/clients were women).[34] Indeed, once statistics in lawful assisted suicide began to be documented in Oregon, roughly one-half of those who sought physician-assisted suicide by prescription were men and one-half were women, which highlights the skewed nature of Kevorkian's practice, a concern for doctors such as Quill; Quill himself noted that data out of Oregon, nearly a decade after his first effort in hospice for his patient, prior to his 1991 willingness to participate in moving the debate forward by publishing his essay about his assisting in the suicide of Diane, "suggest that 'hospice' and 'hemlock' are not mutually exclusive, but in some ways supplement or enhance one another."[35]

Nonetheless, as Quill knew at the time he gave Diane the prescription, and certainly by the time she died, the circumstances of her death exposed both him and her family members to potential criminal liability. The description

Quill gave to the medical examiner regarding Diane's cause of death was, he wrote, "to protect all of us, to protect Diane from an invasion into her past and her body, and to continue to shield society from the knowledge of the degree of suffering that people often undergo in the process of dying."[36] By publishing his essay in the *New England Journal of Medicine*, Quill invited immediate investigation by criminal justice authorities as well as medical administrators. Although there was no data as to how many and how often doctors gave prescriptions or other assistance to terminally ill patients who wanted to commit suicide in the United States, in an interview he gave the week after the publication of the essay, "Dr. Quill said he believed that many doctors had done what he had done but not talked about it and that many other doctors who might be willing to aid a suicide fear to discuss the issue with their patients."[37]

The implicit invitation Quill issued, inviting prosecution of this novel (or new) issue of doctor-assisted suicide (indeed, committed prior to Jack Kevorkian's first hastened death) was taken by Howard R. Relin, the Monroe County district attorney. Relin's initial contemplation was to consider whether, consistent with New York law, Quill should be prosecuted for aiding in a suicide, a crime punishable by up to four years in prison, but noted that "these are very difficult cases because the law is in conflict with people's perception of the right to die."[38] Placed in context, this statement shows that public perception of what they could and could not receive in the scope of medical treatment had changed, and that the role of medicine was no longer simply to cure patients or to implement aggressive palliative care for the sick. With immediacy, Quill's case was contrasted with others, such as an anonymous doctor in residency, who, in 1988, wrote a letter to the editors of the *Journal of the American Medical Association*, entitled, "It's Over, Debbie," in which he acknowledged deliberately giving a lethal injection of morphine to a woman imminently terminally ill with ovarian cancer, a woman he had never met before, and who could barely breathe. A key differentiating factor was that Quill had treated Diane, whereas the resident had not done so with "Debbie," and Kevorkian had not been Janet Adkins's treating physician. Moreover, opponents of euthanasia noted that Quill's presentation was such as to represent an issue whose time for debate had arrived—Dr. Ronald E. Cranford, a neurologist and a medical ethicist associated with the Center for Biomedical Ethics at the University of Minnesota, who told the press, "[t]his is a very important case, and people will have trouble criticizing the procedure."[39] In general terms, these included the longtime doctor-patient relationship, the imminence of the terminal phase of the illness, the additional oncology and hematology consultations, the repeated and enduring requests of Diane, and the psychological consultation, all but predicting what would become legal in Oregon as protocols and regulatory mechanisms by the end of the decade, as voters approved Measure 16. Dr. Cranford, an opponent to euthanasia and physician-assisted suicide, "because of the potential for abuse [i.e., in accord with the arguments of Yale Kamisar in the 1950s], said that a rough measure of the frequency [in 1991] came at a meeting . . . when about 10 percent of the 200 participants raised a hand when asked if they knew directly of a case of active euthanasia."[40] This anonymous and informal survey was carefully worded, by asking if the doctors knew of (rather than implicating themselves by answering a survey as to what they had participated in, underscoring the risk to which Quill had subjected himself). As for Quill himself, he issued the essay about Diane's death, with the permission of the family, because of his reactions to Kevorkian's first case, stating to the press that Kevorkian's first case "focused on machines, and making it a mechanized, sterilized process was not right . . . [he] did not know the person well; that was so far away from anything I [i.e., Quill] could do."[41]

So it was that from the outset, Quill sought to distinguish himself, his practice, and his involvement with Diane's death from that of Kevorkian and Janet Adkins's death, from the outset in 1991. Thus, from the outset, the hospice physician was cast as hero, the former pathologist as antihero. While Kevorkian attained fame and infamy as a folk hero among some of the pro-choice movement (and a folk devil, causing a moral panic, among members of the pro-life movement),[42] and went on to perform over 130 assisted suicides (some of whom were not believed to be terminally ill), Quill pointedly stopped after Diane, and after he brought the issue to public attention. Quill published in the most elite medical journal and with the implicit approval of members of his profession (soon made explicit), whereas Kevorkian was marginalized by (and ultimately excluded from) the medical profession. Quill's ongoing respect for the dignity of Diane and her family were exalted, in stark contrast to the videotapes Kevorkian made of patient-consent colloquies and procedures, themselves the focus of outrage in the criminal justice system and the medical establishment.

That is not to say that Quill did not have experience with the criminal justice system. A grand jury was convened and Quill made the somewhat unusual, and highly risky, choice to testify at that proceeding. After Quill was examined for some three hours, the grand jury declined to indict him. The grand jury had been convened after Monroe County officials, who had been unsuccessful in investigating death records, made the April 1991 discovery of the body of Patricia Diane Trumbull, who died on May 19, 1990, and whose body was being used as an instructional cadaver.[43] Because grand jury records are secret, there is no extant public record of their decision-making process. However, the grand jury could have chosen to indict Quill for manslaughter in the second degree, a class C felony bearing a sentence of 5 to 15 years, and which provides, in pertinent part, "he intentionally causes or aids another person to commit suicide,"[44] Alternatively, the grand jury could have chosen to indict Quill for the crime of promoting a suicide attempt, a class E felony, bearing one to four years in prison, conduct by which "he intentionally causes or aids another person to commit suicide."[45] The grand jury declined to do either, and rather chose to refuse to indict Quill for actions to which he essentially issued a full, clear, and voluntary confession in his essay. What is clear is that, similarly to Yale Kamisar's argument that the law in action provides for remedies that the law on the books would not, the grand jury sought its own remedy of the circumstances presented. Upon learning of this, an opponent to assisted suicide, Dr. Ronald Cranford, nonetheless commented that Quill had behaved "courageously" and that Americans would "support what Dr. Quill did and the way he did it," thus reflecting "a loss of confidence in medicine by the American public and a feeling that they will lose all control over their lives and that their lives will be unduly prolonged."[46] In essence, the fear, as articulated by Quill in his essay, was that "[s]uffering can be lessened to some extent, but in no way eliminated or made benign, by the careful intervention of a competent, caring physician, given current social constraints."[47]

After the grand jury determined that Quill should not be subject to further prosecution, the New York State Health Department also reviewed the case and made a finding that Quill had not engaged in professional misconduct. Indeed, Quill observed that the New York State Health Department went even further, and "recommended that the New York State Task Force on Life and the Law review the overall subject."[48]

Meanwhile, Dr. Nigel Cox, a respected rheumatologist in the United Kingdom, was being investigated for attempted murder with regard to conduct he had engaged in on August 16, 1991, around the same time as the Quill investigation. As with what transpired between Quill and Diane, the essential facts of what Cox did regarding his long-term patient, 70-year-old Lillian Boyes, were not in dispute. Mrs. Boyes was hospitalized with end-stage rheumatoid arthritis, and told doctors at Royal Hampshire County Hospital that she wanted no further treatment, other than painkillers; this was witnessed by Mrs. Boyes's two sons, who maintained a constant vigil over their dying mother.[49] Boyes had suffered from the condition for 30 years and was deemed within hours of death, when she asked for help in dying sooner. Cox unquestionably gave her two ampoules of potassium chloride, enough to be a lethal dose, and with no curative or palliative medical use.[50] Because he had written this down in medical chart notes, a nurse who reviewed the chart felt dutybound to report the matter to hospital administrators, who in turn communicated with the Crown Prosecution Service. The jury found Cox guilty of attempted murder because the body had already been disposed of and an autopsy could not be performed. However, when the jury came in to deliver the verdict, several members openly wept as they convicted.[51] Family members told the press, "the family never made any complaint against him, and nothing has happened [i.e., the jury verdict of guilty] to change our view."[52] Learning of the verdict, Ludovic Kennedy, the vice president of Exit (the voluntary euthanasia society) told members of the press, that "in [his] view, the public shocked by the verdict is now ready for a change in the law. So, for its own protection, is the medical profession."[53]

Whereas a mandatory life sentence may have been necessitated by a body available for autopsy (and thus resulting in a potential conviction of murder), Cox was able to benefit from a discretionary sentence for attempted murder. Mr. Justice Ognall told Cox at sentencing, "the deliberate conduct of a doctor designed to bring about the death of a patient, in my clear judgment, requires as a matter of principle that it should be marked by a sentence of imprisonment."[54] Mr. Justice Ognall imposed a 12-month jail sentence, which it suspended, notwithstanding the fact that the court held that "what [Cox] did was not only criminal, it was a total betrayal of [his] unequivocal duty as a physician."[55] Notwithstanding this "betrayal," Mr. Justice Ognall gave the minimal sentence, which he then suspended, because "[Cox] allowed what [he] knew to be [his] clear duty to be overruled by [his] deep personal distress and compassion for [his] patient, who was on the brink of a painful death."[56] Mr. Justice Ognall went on to allow for extenuating circumstances, mitigating sentence, "[n]o doubt [Cox] did [so] because that utterly splendid lady had become over many years an admired and cherished friend."[57] As an indication that the sentence was case specific, and not to be viewed as a possible indication for how euthanasia cases would be treated, the jurist concluded, that "[i]t is for these reasons that I regard your case as wholly exceptional, if not unique."[58]

Notwithstanding the fact that Cox had been subjected to a winking prosecution, much less severe than the facts would have warranted had an autopsy been available, and the suspended sentence, Cox issued a prepared statement at a news conference, in which he said that he was "devastated by the verdict of guilty and the suspended prison sentence for what was a bona fide act that was solely in the interests of Mrs. Boyes. It seems somewhat harsh to criminalise me for doing my best in what were quite exceptional circumstances."[59] Medical peers did discipline Cox, by requiring him to take palliative care instruction, attend meetings designed "to explore problems and rebuild relationships," and to be supervised for a period of six months;[60] this stood in contrast to Quill, who went undisciplined, as such a disciplinary measure that would have been redundant, had it been imposed upon Cox, a hospice physician who specialized in palliative care.

Unlike Jack Kevorkian, who was the subject of the previous chapter and who engaged in conduct escalating in frequency and in severity of taking control of the end-of-life cases in which he participated, Quill and Cox each engaged in the conduct once. Both were in contexts of long-term patients, and Quill's essay sought to make a single statement to move the debate forward, whereas Cox was widely viewed as one who mistakenly made notes of something he should not have done in the first place, but with no agenda. Indeed, Cox himself said, "I would like to emphasise [*sic*] this was not a trial about the general issue of euthanasia, but about a very specific and most unusual set of circumstances, and I think that is most important [and] my aim now is to get back to as normal a life as possible as quickly as possible."[61] The ways in which Quill and Cox were treated, by both the criminal justice system and the medical administrative boards, stood in sharp contrast to how the state of Michigan viewed Kevorkian's activities, permanently suspending him in 1991 from practice after the Wantz/Miller assisted suicides and creating a law prohibiting assisted suicide, which was widely believed to be designed specifically to stop him.

LEGISLATIVE EFFORTS AND LITIGATION SUBSEQUENT TO THE DOCTOR PROSECUTIONS OF THE 1990s

Like the doctor prosecutions of the 1950s, these cases served to stimulate public interest in a debate that in the 1990s had a new concept—that of physician-assisted suicide. However, these cases also stimulated discussion of euthanasia, including whether there would be a slippery slope from voluntary assisted suicide to non-voluntary or involuntary euthanasia, or to euthanasia of vulnerable populations such as the elderly and minorities, who might be subject to economic pressure. In addition, concern began to grow that a disproportionate number of women and younger people with chronic, debilitating illnesses, particularly in view of the advent of AIDS,[62] would be either motivated to seek assisted suicide, or somehow pressured into it, by virtue of being in a vulnerable population of what was (at that time) primarily an illness afflicting members of the gay community and/or intravenous drug users.

First, and specifically in response to Quill's prosecution in New York, the New York State Task Force on Life and the Law convened to consider whether to recommend changes in the law regarding assisted suicide, consistent with its "broad mandate to recommend public policy on issues raised by medical advances. . . . Assisted suicide and euthanasia were not on the agenda initially presented to the Task Force [by Governor Mario Cuomo in 1984]. Nor was the prospect of legalizing the practices even remotely part of public consideration at that time."[63] The Task Force noted that "although no major efforts to legalize assisted suicide and euthanasia have been launched in New York State, [they] chose to examine the practices and to release this report in order to contribute to the debate unfolding in New York and nationally."[64] The Task Force ultimately determined—unanimously—that the dangers of a dramatic change in law and policy regarding assisted suicide would "far outweigh any possible benefits" that might adhere to being legally able to choose the time and manner of a terminally ill person's death; in specific, the Task Force held that "the risks would be most severe for those who are elderly, poor, socially disadvantaged or without access to good medical care."[65] However, the Task Force, which was composed of doctors, lawyers, nurses, clergy, and medical ethicists, did make recommendations for improving palliative care,[66] diagnosing and treating depression associated with chronic and terminal illness,[67] and offering additional social and other services as a response to those patients who request euthanasia or assisted suicide.[68] These findings paralleled those of the House of Lords Select Committee on Medical Ethics, which convened from 1993 to 1994, and issued, similarly to the New York State Task Force on Life and the Law, a recommendation that the law go essentially unchanged, but recommended that further attention and development be made in the area of palliative care.[69]

Second, and on the other side of the country, the combination of no indictment against Quill and the publication of Hemlock president Derek Humphry's controversial book, *Final Exit: The Practicalities of Self-Deliverance and Assisted Suicide for the Dying*,[70] prompted a Universalist Unitarian minister, Ralph Merro, to found and direct a pro-initiative campaign for a legislative initiative allowing for lawful physician-assisted suicide for terminally ill patients.[71] Developing and promoting the initiative became part of Hemlock's national agenda.

Washington I-119 was drafted with three basic provisions; first, to define "persistent vegetative state" and "irreversible coma" as conditions under which life-support systems could be withdrawn if a patient requested such withdrawal in a living will.[72] The second provision listed artificial nutrition and hydration as life-sustaining procedures which could be refused or withdrawn.[73]

It was, however, the third provision of I-119 that proved to be controversial. This provision "stated that physicians could provide assisted suicide and active euthanasia as medical services to competent, terminally ill adult patients who request them. Dubbed physician 'aid in dying' by the initiative's sponsors, this last provision would [have made] Washington state the only place in the world where the intentional killing of patients by physicians had been formally decriminalized."[74] Initiative 119 was voted upon barely days after Kevorkian assisted in the October 23, 1991, deaths of Marjorie Wantz (who suffered from chronic female-specific pain) and Sherry Miller (who had multiple sclerosis), and Washington voters rejected Initiative 119 by a vote of 54 percent to 45 percent, notwithstanding a list of safeguards and regulatory mechanisms for any aid in dying.[75] Despite its defeat, Washington State Initiative 119 has a place in history as the first public initiative to allow for lawful physician-assisted death for mentally competent patients with certifiably terminal conditions,[76] Although Kevorkian's activities placed the initiative's potential for success at risk, and was, in effect a free advertisement for the opposition (which included the Washington State Medical Association and Catholic churches and bishops around the country),[77] the relatively slim majority showed that the issue was one whose time had come for debate, and its very development was historic. During the following election cycle, euthanasia proponents "offered a similar initiative for signatures for the ballot in California, a state with a tradition of common and protracted use of the initiative process. The players would be similar to the players in Washington and the result would be the same."[78] As an irony, the one state that Kevorkian had been licensed in, other than Michigan, had been California, and while he lost his license there, as well as in Michigan, in 1991, one prong of spirited attack by pro-life advocates was that Kevorkian might seek to set up "death clinics" in California, should Initiative 161 pass. This was a reasonable argument, given that Kevorkian had continued, and indeed, accelerated, his practice between 1991 and 1992. So it was that California voters rejected Initiative 161, which had sought to legalize both euthanasia (by lethal injection) and assisted suicide, by a margin of 54 percent to 46 percent.[79] Up until (and indeed, the day after) the election, it was uncertain whether the initiative would pass. However, exit polls showed that women, older voters, Asians, and African Americans were most likely to vote against the measure, and that the exit polls showed those in favour of the initiative were most likely to be male, under 30, and with a stated income of over $75,000 (1992 dollars).[80] In other words, those voting against the measure were most likely to be minorities or members of vulnerable populations. In addition, there was a strong religious opposition—one telephone survey conducted by the Terrance Group, showed that 60 percent of Catholic voters, 55 percent of Protestant voters, higher percentages among Baptists, fundamentalists, and Pentecostalists, and 81 percent of regular churchgoers were in opposition to the initiative.[81]

During the mid-1990s, in the wake of the defeat of Washington Initiative 119, and the prosecution (albeit unsuccessful) of Quill, a group called

Compassion in Dying set a pair of test cases, in the state originating the Quill grand jury (New York) and in the state originating the first early public initiative (Washington). Almost simultaneously, euthanasia and assisted-suicide advocates sought to put another initiative on the ballot, this time in the state of Oregon. The 1994 Oregon initiative, the subsequent (failed) recall effort, litigation, and implementation will be treated in the next chapter, as this regards a new legal mechanism for (and litigated into) a new millennium.

However, and also commencing in 1994, a group of physicians (including Quill) and terminally ill patients field suit for seeking injunctive relief from state and local officials from enforcing New York Penal Law Sections 125.15 (manslaughter in the second degree) and 120.30 (promoting a suicide attempt), to the extent that they criminalized physician-assisted suicide. Because patients in the original case had died, the physicians proceeded without them.[82] At the same time, Compassion in Dying (a nonprofit group), physicians, and three individual terminally ill patients, sought a declaratory judgment in federal court to similarly strike down Washington Rev. Code Section 9A.36.060 (promoting a suicide attempt).[83] Both cases were a matter of first impression on the constitutionality of the respective states' criminal prohibitions against physician-assisted suicide.

Contextually pertinent, as well as historical, is how the cases were constructed—each of the suits had both doctor and patient plaintiffs, seeking judicial declaration that criminal provisions prohibiting physician-assisted suicide were unconstitutional. Each of the physicians had involvement with patients facing terminal illnesses (in contrast to Kevorkian, who had been a pathologist). In New York, Quill filed a declaration with the court attesting to his "very public criminal investigation," whereas the other two physicians attested to requests from terminally ill patients who sought assisted suicide, which the physicians declined due to fear of prosecution under the New York statute.[84]

In Washington State, none of the doctors had ever been prosecuted, but all attested to fear of criminal liability. Harold Glucksberg was a cancer specialist and on the clinical faculty of the University of Washington School of Medicine, Abigail Halpern was the medical director of a family practice with patients dying of AIDS, Thomas Preston was a cardiologist who attested that he treated patients whose deaths he believed he should hasten (but did not, for fear of the criminal law), and Peter Shalit was a family practitioner and the medical director of the Seattle Gay Clinic.[85] As an initial matter, both cases were composed of esteemed medical professionals working, teaching, and writing within the medical establishment, and hence with support that Kevorkian did not have from the profession that had ejected him from practice.

The patient plaintiffs added additional historical texture to the New York and Washington cases. The New York patient plaintiffs and the doctor plaintiffs each alleged in the original complaint that they "were mentally competent adults; that they were in the terminal stages of fatal diseases; that they faced progressive loss of bodily function and integrity as well as increased suffering; and that they desired medical assistance in the form of medications prescribed by physicians to be self-administered for the purpose of hastening death."[86] The individual (and anonymous) patient plaintiffs in the Washington litigation were a 69-year-old female physician and cancer patient (who died prior to the first district court decision), a 44-year-old male artist and end-stage AIDS patient (already blinded by the illness at the time of filing, and deceased by the time of the Ninth Circuit decision), and a 69-year-old man who was constantly connected to an oxygen tank prior to his death pending appeal in the Ninth Circuit.[87] These patients were as compelling in their lucidity as in their factually sympathetic illnesses. In other words, in a few short years, cases went from being based upon private family statements that they did not want to end up like Karen Ann Quinlan (or Nancy Cruzan) to being based upon public declarations of educated and activistic patients, who gave shape and context to the historical arguments they were making and creating. While both cases were unsuccessful at the Supreme Court level, they each met with profound measures of success on the Circuit Court federal appeals level before those final losses.

For example, in *Quill v. Vacco*, the Second Circuit Court of Appeals held, insofar as the New York case goes, that:

> 1) the statutes in question fall within the category of social welfare legislation and therefore are subject to rational basis scrutiny upon judicial review; 2) New York law does not treat equally all competent persons who are in the final stages of fatal illness and wish to hasten their deaths; 3) the distinctions made by New York law with regard to such persons do not further any legitimate state purpose; and 4) accordingly, to the extent that the statutes in question prohibit persons in the final stages of terminal illness from having assistance in ending their lives by the use of self-administered, prescribed drugs, the statutes lack any rational basis and are violative of the Equal Protection Clause.[88]

The Ninth Circuit Court of Appeals, in a rare *en banc* (full judicial panel, rather than a three-judge seating) review, also held that terminally ill patients should be able to determine the time and manner of their deaths, as a matter of a constitutionally protected liberty interest under the due process clause.[89] As a curious coincidence, this decision was released on March 6, 1996, while Kevorkian was on trial for two assisted suicides, of which he was acquitted on March 8, 1996.

Further, although the U.S. Supreme Court unanimously ruled that there is no constitutional right to assisted suicide, the members of the bench left it to individual states to pass laws regarding and, where appropriate, regulating assisted suicide.[90] The Supreme Court so declared by the paired decisions of *Washington v. Glucksberg*[91] and *Quill v. Vacco*.[92] In *Glucksberg*, the court held that there was no right to have (or to provide) physician-assisted suicide under the due process clause; Glucksberg was himself a physician (who was joined by four other doctors, as well as several terminally ill patients and Compassion in Dying). In *Quill* (which was physician-focused in its construction), the court held that the New York statute, under which Quill had been briefly prosecuted, did not violate the equal protection clause of the Fourteenth Amendment.

While the court was unanimous in its holding, there were a variety of concurring writings. Perhaps the most interesting of these was that of Justice Sandra Day O'Connor, who wrote that notwithstanding her agreement in the court's decisions, she also noted that everyone would either be subject to, or have family member(s) subjected to the sort of illness that might result in a decision to seek assisted suicide, and that states would individually have to decide how to face this. This commentary had a backstory that would prove to be common with many (if not most) involved with promoting the assisted-suicide debate during the 1990s—profound personal experience. While not written about in her concurring writing, it was common knowledge that she herself was a cancer survivor, having successfully battled breast cancer in 1988.

O'Connor is not alone in having had close personal (or family) experience. Judge Breck, who presided over the second pair of 1996 Kevorkian trials (for the Wantz/Miller deaths) spoke at length about the death trajectory of his wife, which unquestionably influenced (favorably) how he felt about physician-assisted suicide. Thus, the compelling state interests of preservation of human life and protection of those who were not capable of informed consent, traditional and continued by the *Glucksberg* and *Quill* cases, were beginning to show potential mitigation (or, pro-life advocates might say, erosion) when taken together with life experiences of jurists. This social construction of context, outside of the arena of the law, yet alluded to by both federal and state jurists, should not be discounted. Indeed, it may perhaps have laid the groundwork for the next decade, when the U.S. Supreme Court would effectively dodge the profound questions regarding the right (or not) to assisted suicide (or to provide it) when the Oregon legislation was challenged, and also went to the U.S. Supreme Court (where it was upheld, on largely procedural grounds).

CONCLUSION: DOCTOR PROSECUTIONS AT THE END OF A DECADE OF DEBATE

Perhaps the most interesting point to be made at the close of the 1990s was that what had appeared to be a blurry and academic line of whether and how euthanasia could be abused in the second half of the century, and the definitions and risks of physician-assisted suicide, became much clearer as the decade ended. Jack Kevorkian had been incarcerated in Michigan and had become a poster child for the potential for abuse, as an out-of-license doctor, who had no patients other than those who sought death, and who accelerated his practice to flout the law until the media darling came to be demonized by his own publicity-seeking behavior. On the other side of the Atlantic Ocean, Harold Shipman was under investigation for 15 murders by euthanasia by lethal injection, of 80 percent female patients, a gender differential similar to that of Kevorkian; in the year 2000, he would be sentenced to life in prison, and an inquiry would suggest that he had more likely hastened the deaths of some 250 patients, without their knowledge. Timothy Quill, the paradigmatic hospice doctor who academically and legally brought assisted suicide into the public sphere by writing about Diane, continued to practice, teach, and write advocating choices in dying. While there are anonymous admissions of death-hastening conduct, these do not find their way into the public sphere outside of Oregon, where there is a highly regulated practice that will be the subject of the next chapter.

The historical context of doctors was not the only matter changed in the euthanasia and assisted-suicide debate. The 1990s saw the rise of patient activism, with their doctors and with the legal system. The family of Nancy Cruzan started the decade with a single case about a single patient at the beginning of the decade, but patients themselves began to take their battles to the voting booth and to the courthouse. The advent of AIDS may have changed some of the dynamics of these battles.

NOTES

1. 497 U.S. 261 (1990).
2. *Id.*, 269–85.
3. Statement of Barbara Walters, "Hot Topics" segment, *The View*, ABC, June 6, 2011.
4. Post, Stephen G., "A Postmortem on Initiative 119," *Health Progress*, January–February 1992, 70.
5. 521 U.S. 702 (1997).
6. 521 U.S. 793 (1997).
7. *People v. Kevorkian*, sentencing proceeding, April 13, 1999, 37–38.

8. No. DA 09-0051 (Mont. Sup. Ct. December 21, 2009).

9. Quill, Timothy E., "Sounding Board, Death and Dignity: A Case of Individualized Decision Making," *New England Journal of Medicine* 325, no. 10 (March 7, 1991): 691–94.

10. *Id.*

11. *Id.*

12. *Id.*

13. *Id.*

14. *Id.*

15. *Id.*

16. *Id.*

17. *Id.*

18. Hillyard, Daniel, and John Dombrink, *Dying Right: The Death and Dignity Movement* (New York: Routledge, 2001).

19. Quill, *op. cit.*, 691–94.

20. *Id.*

21. *Id.*

22. Swarte, N. V., M. I. van der Lee, J. G. van der Boom, J. Van den Bout, and A. P. Heinz, "Effects of Euthanasia on the Bereaved Family and Friends: A Cross-Sectional Study," *British Medical Journal* 327, no. 7408 (July 26, 2003): 189.

23. *Id.*, 189, fn. 4.

24. Quill, 1991, *op. cit.*

25. *Id.*

26. *Id.*

27. *Id.*

28. *Id.*

29. *Id.*

30. *Id.*

31. *Id.*

32. Altman, Lawrence K., "Jury Declines to Indict a Doctor Who Said he Aided in a Suicide," *New York Times*, July 27, 1991.

33. *Id.*

34. Dragovic, Ljubisa J., et al., "Dr. Kevorkian and Cases of Euthanasia in Oakland County, Michigan, 1990–2000," *New England Journal of Medicine* (1998): 1735–36.

35. Timothy Quill, "Foreword," in Constance E. Putnam, *Hospice or Hemlock: Searching for Heroic Compassion* (Westport, CT: Praeger, 2002), p. xii.

36. Quill, 1991, *op. cit.*

37. Altman, Lawrence K., "Doctor Says He Gave Patient Drug to Help Her Commit Suicide," *New York Times*, March 7, 1991.

38. *Id.*

39. *Id.*

40. *Id.*

41. *Id.*

42. Cohen, Stanley, *Folk Devils and Moral Panics: The Creation of the Mods and the Rockers*, 2nd ed. (Oxford, England: Blackwell Publications, 1987).
43. Altman, Lawrence K., "Jury Declines to Indict a Doctor Who Said He Aided in a Suicide," *New York Times*, July 27, 1991.
44. New York Penal Law Section 125.15 (3).
45. New York Penal Law Section 120.30.
46. Altman, July 27, 1991, *op. cit.*
47. Quill, 1991, *op. cit.*
48. Quill, Timothy, *Death and Dignity: Making Choices and Taking Charge* (New York: W.W. Norton & Company, Inc., 1993), 21.
49. Mullin, John, "Widow's Son Backs Consultant in Lethal Injection Case," *Guardian* (London), September 21, 1992.
50. *Id.*
51. *Id.*
52. *Id.*
53. *Id.*
54. Mullin, John, "Lethal Injection Doctor Given Suspended Sentence," *Guardian* (London), September 22, 1992.
55. *Id.*
56. *Id.*
57. *Id.*
58. *Id.*
59. *Id.*
60. Dyer, Clare, "Convicted Doctor to Take Up Old Job," *Guardian* (London), December 4, 1992.
61. Mullin, September 22, 1992, *op. cit.*
62. Ogden, Russell D., *Euthanasia and Assisted Suicide in Persons with Acquired Immunodeficiency Syndrome (AIDS) or Human Immunodeficiency Virus (HIV)* (New Westminster, British Columbia: Peroglyphics Publishing, 1994).
63. New York State Task Force on Life and the Law, *When Death Is Sought: Assisted Suicide and Euthanasia in Context* (self-published: May 1994), Preface, vii.
64. *Id.*
65. *Id.*, ix.
66. *Id.*, 153–58.
67. *Id.*, 175–77.
68. *Id.*, 177–81.
69. House of Lords, *Report of the Select Committee on Medical Ethics* (HL Paper 21), vol. 1 (Report) (London: HMSO).
70. *Final Exit: The Practicalities of Self-Deliverance and Assisted Suicide for the Dying* (Eugene, Oregon: Hemlock Society, 1991).
71. Hillyard, Daniel, and John Dombrink, *Dying Right: The Death with Dignity Movement* (New York: Routledge Publications, 2001), 33.
72. *Id.*
73. *Id.*

74. *Id.*, 35.
75. *Id.*, 36.
76. McGough, Peter M., "Washington State Initiative 119: The First Public Vote on Legalizing Physician-Assisted Death," *Cambridge Quarterly of Healthcare Ethics* 2 (1992): 63–67.
77. Hillyard and Dombrink, *op. cit.*, 45–52.
78. *Id.*, 53.
79. Steinfels, Peter, "Help for the Helping Hands in Death," *New York Times*, February 14, 1991, A1. (ProQuest).
80. *Id.*
81. *Id.*
82. *Quill v. Koppell*, 870 F. Supp. 78 (S.D.N.Y. 1994).
83. *Compassion in Dying v. State*, 850 F. Supp. 1454 (W.D.Wa. 1994).
84. 870 F. Supp. 78.
85. *Compassion in Dying v. State of Washington*, 49 F.3d 586, 589 (9th Cir. 1995).
86. 870 F. Supp. at 79–80.
87. 49 F.3d. at 588.
88. 80 F.3d 716, *op. cit.*
89. 79 F.3d 790.
90. *Washington v. Glucksberg*, 521 U.S. 702 (1997).
91. *Id.*
92. *Vacco v. Quill*, 521 U.S. 793 (1997).

7

Legalizing and Implementing Physician-Assisted Suicide in Oregon in the 1990s and 2000s

INTRODUCTION

While the American physician prosecutions of Dr. Kevorkian in Michigan and Dr. Quill in New York led criminal prosecutions (successful and not), criminal legislation (in Michigan), and the unsuccessful civil litigation to declare the New York (and Washington State) laws constitutionally invalid, the citizens of Oregon chose a different path of contemplating the assisted-suicide debate. This was even as Michigan was criminalizing assisted suicide, and Washington State and California were sustaining electoral losses in efforts to decriminalize medical euthanasia. While New York and Washington State faced civil litigation challenging their statutes that prevented physician-assisted suicide, Oregon was preparing to sponsor a voter initiative for the November 1994 election. This initiative was, however, for physician-assisted suicide, rather than medical euthanasia.

Technically, the movement for a legislative initiative in Oregon was not preceded by a doctor being prosecuted there. However, Janet Adkins, Jack Kevorkian's first patient, was an Oregonian. Thus, in a very real way, Oregon was also ground zero for the 1990s assisted-suicide debate and the Kevorkian cases—as taken from the patient's perspective. While doctor prosecutions by definition stimulated much of the legal debate, patients in most of those cases were (prior to the Kevorkian cases and the related Hobbins case) nonpolitical. In effect, families of patients (as in the Quill case of the 1970s and the Cruzan case of the late 1980s and decided by the U.S. Supreme Court in 1991) were forced into action because of loved ones in persistent vegetative states, rather than terminally ill (but still quite alive) family members. These terminally ill patients were seeking to have their own voices heard, their own choices made by themselves, and their own decisions about the time and the manner of their deaths put into effect. Those patients who did not want to pursue assisted suicide did not have to act at all.

Furthermore, Oregon was an initiative-friendly state, which is to say that placing voter initiatives on the ballot was a relatively common practice and provided an alternative to civil litigation and criminal prosecution. In the election cycles preceding the 1994 placement of Measure 16, the Oregon Death with Dignity Act, on the ballot, there was a measure for term limits (which passed in 1992), a measure to limit the rights of gays and lesbians (which failed in 1992), a proposal to ban abortion (which failed in 1990), and a proposal to require parental notification for abortions (which failed in 1992); these were outcomes that led "proponents [of assisted death legislation to] speculate that choice would be a winning theme for Oregonians,"[1] and that so-called choice in dying by physician-assisted suicide would be likely to pass by ballot initiative. Thus, rather than promote social and legal change by a test balloon case of a doctor who engaged in, and essentially confessed to, committing a criminal offense, or the alternative process of suing the state for a declaration invalidating the law, the assisted-suicide debate was moved forward in Oregon by a grassroots movement for legislative change. The Michigan morass as the Kevorkian cases developed, the New York non-indictment in the Quill case, and the physician/patient suits for declaratory judgment in Michigan, New York, and Washington pointed to the ballot initiative as a potentially more satisfying way of debating the issue and effecting legal change regarding the "criminality" of assisted suicide, with all euthanasia alternatives cast aside as unsuitable in the Oregon effort.

Oregon Right to Die, the political action committee that sponsored the Oregon Death with Dignity Act, began soliciting campaign contributions to fund a petition and signature collection five months before the general election; in its initial mailing, director Eli Stutsman included a copy of the act, a summary of its provisions, related press materials, and literature.[2] Giving additional historical context to the way in which the Oregon initiative was constructed, Oregonians had the benefit of a valuable lesson from the recently failed Washington and California ballots, and sought to create a law which legalized assisted suicide by prescription for oral medication, rather than medical euthanasia by injection or other means. In addition, the Oregon effort sought consensus building by way of its chief petitioners. Barbara Coombs Lee had been a nurse turned attorney, and Dr. Peter Goodwin was a physician who taught family medicine. In other words, these were esteemed members of the medical community much like Quill, rather than outcasts like Kevorkian. Thus, the heart of the team drafting the legislation was inclusive of medical and legal professionals, who distanced themselves from fellow Oregonian Derek Humphry, who authored *Final Exit* (a how-to book on suicide) and had previously founded the Hemlock Society, a right-to-die advocacy group. Like Kevorkian, Humphry was widely viewed as extremist,[3] and as one who would view assisted suicide to be a pathway to euthanasia—in other words, to be the personification of Kamisar's wedge to the slippery slope.

Opposition of Measure 16 (the provisions of which will be discussed later in this chapter) included the Oregon Pharmacists Association, which protested to the effect that the role of pharmacists was to promote and prolong life, rather than to deliberately shorten or terminate it. Likewise, Oregon Hospice opposed the measure, and sought for patients to seek palliative care and hospice. However, these are not necessarily inconsistent, as the case of Quill demonstrated, and as the future of Oregon, post-legislation would also statistically show. Thus, it was not only religious groups (primarily Catholic organizations in terms of advertising efforts) who opposed the legislation and legalization of physician-assisted suicide, by any name, but also professional groups who were closely linked to the medical care of the dying.

Measure 16, the first voter initiative that would permit legal assisted suicide, succeeded where prior initiatives for lawful medical euthanasia failed. (However, the initiative process would not prove to be equally successful in other states, as the 1998 experience of Michigan's Merian's Friends would discover later). On November 3, 1994, Oregonians passed the measure, with 51.3 percent of those who turned out voting in favor and 48.7 percent voting in opposition.[4]

Oregon had, in addition to a strong team, procedural advantages that made the state friendly to the concept of a voter initiative. It was the first state to adopt an initiative process, and "citizen lawmaking has often been used in grand social movements,"[5] such as adopting women's suffrage by initiative, decriminalizing (albeit temporarily) marijuana possession and use, and passing same-sex-friendly laws. A socially progressive and relatively secular state, Measure 16 was more limited in scope than the Washington and California reform efforts, providing for physician participation only to the extent of providing a prescription for barbiturates for the patient to fill, then use at the patient's own discretion. In other words, Measure 16 emulated the model of New York Dr. Timothy Quill and eschewed Kevorkian as an exemplar.

SURVIVING CHALLENGES TO THE OREGON LAW

Passage of the Oregon measure was not the end of the debate. First, there was a lawsuit prior to implementation of the Oregon law, *Lee v. Oregon*,[6] whereby plaintiffs (a collective of doctors, patients, and medical facilities) sought to prevent the Oregon law from being put into effect. During the pendency of this case, a preliminary injunction was in place, as a result of which the law was not implemented until 1997.

Second, there was a recall effort to repeal the Oregon Death with Dignity Act, whereby the House and Senate of the Oregon legislature returned Measure 16 to the voters by Measure 51 in a special election; rather than repeal the law, Oregon voters chose to reaffirm the measure, by a margin of 60 percent to 40 percent, in 1997.[7] In other words, the Oregon Death with Dignity Act not only survived the debate, but picked up an additional 10-point margin (or, approximately, a 20 percent bump on top of the proportion of voters who originally voted in favor of Measure 16).

Third, there was a U.S. Supreme Court case, *Gonzales v. Oregon*,[8] in which the Supreme Court held that the U.S. attorney general (originally John Ashcroft) could not use the federal Controlled Substances Act as a mechanism against physicians who duly prescribed pharmaceuticals in accord with the Oregon Death with Dignity Act. This emanated from a 2001 Interpretive Ruling by then Attorney General Ashcroft, to the effect that, because the drugs that Oregon physicians prescribe under the Oregon Death with Dignity Act are regulated by the Federal Controlled Substances Act,[9] physicians who prescribed medication under the Oregon Death with Dignity Act would be in violation of the federal law. While the Supreme Court did not question the power of the federal government to regulate drugs, it held that it was improper to determine how states could regulate pharmaceuticals. While the majority focused upon procedural law, the dissent by Justice Scalia addressed the substantive issue, in which he expressed the opinion that a prescription designed to result in death was not a legitimate medical purpose.

Thus, the U.S. Supreme Court, in its decision in *Gonzales v. Oregon*,[10] in effect rounded out a trio of assisted-suicide decisions emanating from events in the 1990s. Having refused to hear any aspect of the Kevorkian cases was effectively a decision that the state of Michigan was appropriate in its legislative responses and prosecutorial efforts against Dr. Death. Having issued multiple writings in the 1997 decisions arising out of New York and Washington State, the court effectively expressed the collective opinion that there was no federal equal protection and no federal due process right to physician-assisted suicide; however, the various opinions of the justices themselves indicated that there might be circumstances under which would be held to be proper, assuming that it was lawful as a matter of state law. The 2006 Oregon case presented the opportunity to determine whether lawful physician-assisted suicide, pursuant to state law should—or should not—be upheld and permitted. This opportunity the majority declined, choosing instead to support an opinion that chastised the attorney general for a "power grab."[11] From a historical perspective, the Supreme Court in effect carried on forward from the 1997 decisions leaving the question of whether and how to legalize (or criminalize) assisted suicide to the states.

So it was that Measure 16 and the Oregon Death with Dignity Law survived civil challenge, a legislative effort for repeal, and a federal effort to nullify by criminalization.

THE OREGON DEATH WITH DIGNITY ACT AND ITS IMPLEMENTATION

Taken as a given was that the law sought to address concerns regarding doctor-patient communication, the risk of misdiagnosis or abuse, and issues of voluntariness in allowing for physician-assisted suicide. The effective date of the Oregon Death with Dignity Law was October 27, 1997. Starting on that day, Oregonians who were terminally ill had the lawful opportunity to self-administer lethal doses of medication that were prescribed by their physician for that expressed purpose. In other words, the very conduct of which Dr. Timothy Quill wrote of and was briefly (albeit unsuccessfully) prosecuted for, and unsuccessfully sued the state of New York to have declared legally permissible, became perfectly legal under Oregon law.

By perfectly legal, that is to say, an Oregon patient similarly situated to that of Quill's patient, and a physician who, like Quill, wrote a prescription for a lethal dose of pharmaceuticals, would be deemed to have engaged in lawful conduct if the litany of procedural parameters of the Oregon law were followed. By expressly stating the protocols to be followed for an assisted suicide, all other actions and activities that did not adhere to the protocols effectively became criminalized. In effect, by applying the doctrine of *expressio unius exlcusio alterius* (the express inclusion of one excludes the others not mentioned), acts (such as a lethal injection) or *actors* (such as Kevorkian, for being unlicensed) became illegal and thereby prosecutable, by virtue of the Oregon Death with Dignity Law. Each of the provisions will now be enumerated, with brief commentary regarding inclusions and exclusions.

Oregon Revised Statutes (or "ORS") 127.800 Section 1.01 provided for "definitions" of words and phrases under the Death with Dignity Act. In essence, this was important because many of the terms and phrases bandied about by both pro-life and pro-choice advocates may have the potential for vagueness, and vagueness can cause a law to be invalidated, if challenged. Definitions also established boundaries for the parameters to be set forth. With specificity, the statute provided that:

The following words and phrases, whenever used in ORS 127.800 to 127.897, have the following meanings:

(1) "Adult" means an individual who is 18 years of age or older.
(2) "Attending physician" means the physician who has primary responsibility for the care of the patient and treatment of the patient's terminal disease.
(3) "Capable" means that in the opinion of a court or in the opinion of the patient's attending physician or consulting physician, psychiatrist or psychologist, a patient has the ability to make and communicate health care decisions to health care providers, including communication through persons familiar with the patient's manner of communicating if those persons are available.
(4) "Consulting physician" means a physician who is qualified by specialty or experience to make a professional diagnosis and prognosis regarding the patient's disease.
(5) "Counseling" means one or more consultations as necessary between a state licensed psychiatrist or psychologist and a patient for the purpose of determining that the patient is capable and not suffering from a psychiatric or psychological disorder or depression causing impaired judgment.
(6) "Health care provider" means a person licensed, certified or otherwise authorized or permitted by the law of this state to administer health care or dispense medication in the ordinary course of business or practice of a profession, and includes a health care facility.
(7) "Informed decision" means a decision by a qualified patient, to request and obtain a prescription to end his or her life in a humane and dignified manner, that is based on an appreciation of the relevant facts and after being fully informed by the attending physician of:
    (a) His or her medical diagnosis;
    (b) His or her prognosis;
    (c) The potential risks associated with taking the medication to be prescribed;
    (d) The probable result of taking the medication to be prescribed; and
    (e) The feasible alternatives, including, but not limited to, comfort care, hospice care and pain control.
(8) "Medically confirmed" means the medical opinion of the attending physician has been confirmed by a consulting physician who has examined the patient and the patient's relevant medical records.
(9) "Patient" means a person who is under the care of a physician.
(10) "Physician" means a doctor of medicine or osteopathy licensed to practice medicine by the Board of Medical Examiners for the State of Oregon.
(11) "Qualified patient" means a capable adult who is a resident of Oregon and has satisfied the requirements of ORS 127.800 to 127.897 in order to obtain a prescription for medication to end his or her life in a humane and dignified manner.
(12) "Terminal disease" means an incurable and irreversible disease that has been medically confirmed and will, within reasonable medical judgment, produce death within six months. [1995 c.3 s.1.01; 1999 c.423 s.1]

By these definitions, a number of immediate issues were removed from debate. For example, Oregonian Janet Adkins, who had been Kevorkian's first case, would not have qualified under the Oregon law, as her early-onset Alzheimer's was not, within reasonable medical judgment, going to result in death within six months, a procedural requirement. Indeed, film clips of Adkins made in the weeks before her death, and her interview with Jack Kevorkian before he assisted her suicide, showed a woman who was conversant, lucid, and able-bodied (who reportedly played tennis in the weeks before her death, a game that requires physical and mental stamina as well as intellectual planning and strategizing). Kevorkian himself would have been excluded from the Oregon law (even prior to having his license stripped), given his prior area of expertise was in pathology (and that he was not a treating or consulting physician), and therefore procedurally excluded. Furthermore, after the Wantz/Miller cases (the second and third of Kevorkian's assisted suicides), Kevorkian would have been excluded from practice under the Oregon law initially by not having an Oregon license, but also by virtue of loss of licensure, which removed him from the definitional parameters of "health care provider."

The next set of provisions regarded "written request for medication to end one's life in a humane and dignified manner," i.e., by assisted suicide. Thus, Section 2, ORS 127.805 s.2.01, as to "who may initiate a written request for medication," provided that:

> (1) An adult who is capable, is a resident of Oregon, and has been determined by the attending physician and consulting physician to be suffering from a terminal disease, and who has voluntarily expressed his or her wish to die, may make a written request for medication for the purpose of ending his or her life in a humane and dignified manner in accordance with ORS 127.800 to 127.897.

These "voluntariness" clauses went on to be delineated further in the statute, with specificity. However, what is important at this juncture is to note that to qualify for death with dignity (assisted suicide, not euthanasia), required an adult, capacity, residency, terminal illness, attending and consulting physicians, a voluntary request, and compliance with each of the provisions within the statute. These were stated in the conjunctive; in other words, each standard had to be met. Although the sum of these considerable parts would not suffice to surmount the objections of those who take a theological pro-life approach in opposition to assisted suicide, they went very far in assuaging the concerns of those who take a "non-religious" approach objecting to assisted suicide, such as those emanating from Yale Kamisar and his progeny, regarding the thin edge of a wedge onto a slippery slope to non-voluntary, then involuntary, euthanasia. In addition, while the state has an interest in the protection of human life, the purpose of the Oregon Death with Dignity Act was to allow for exclusion for terminally ill patients to be able to personally engage in self-determination.

The statute continued:

> (2) No person shall qualify under the provisions of ORS 127.800 to 127.897 solely because of age or disability. [1995 c.3 s.2.01; 1999 c.423 s.2]

Perhaps the single most important word in this provision is "solely." This provision in effect intones against disposing of the aged (a concern as life spans grow) or of the disabled (historically a concern because of the Nazi experience, contemporarily a concern because of the emergence of long-term degenerative illnesses, such as ALS, MS, Huntington's disease, and Alzheimer's).

The "form of the written request" is addressed by ORS 127.810 s.2.02. As a general comment, procedural rules (relating as to how to do something) can be as important as those of a substantive nature (relating to why to do something). In the Oregon Death with Dignity Act, the form of the written request requires the following elements:

> (1) A valid request for medication under ORS 127.800 to 127.897 shall be in substantially the form described in ORS 127.897, signed and dated by the patient and witnessed by at least two individuals who, in the presence of the patient, attest that to the best of their knowledge and belief the patient is capable, acting voluntarily, and is not being coerced to sign the request.

This summary requirement of the form of the request was also stated in the conjunctive—requiring that each and every element be met. Thus, as to physician-assisted suicide in Oregon, the request signed and dated required (by definition) the ability of the patient to write, plus there must be two witnesses at the time, plus they must have a good faith belief of competency, plus they must have a belief (or knowledge) that the patient is not being coerced or under duress to seek assisted suicide. Again, these were cumulative protocols of protection to ensure voluntariness on the part of the patient, which were legislatively offered to ensure against the thin edge of Kamisar's wedge from which there may emerge a slippery slope to non-voluntary or involuntary assisted suicide.

> (2) One of the witnesses shall be a person who is not:
> (a) A relative of the patient by blood, marriage or adoption;

This provision would protect against a family, with the vested interest of either inheritance or the burden of care of the patient, from being in the position of pushing a vulnerable patient toward an unwanted death.

> (b) A person who at the time the request is signed would be entitled to any portion of the estate of the qualified patient upon death under any will or by operation of law;

This provision sought to protect against those who might have an economic motivation in the death of the patient, whether a relative or a spiritual descendant of Dr. John Bodkin Adams. In the alternative, there was an "or" following subsection (b), prohibiting as a witness:

> (c) An owner, operator or employee of a health care facility where the qualified patient is receiving medical treatment or is a resident.

By this provision, the law protected against facilities becoming glorified death camps and also against individuals who may have a political or economic agenda, who are employed by a facility.

The next requirement of the written form of the request would not only have placed Kevorkian outside the ambit of the law (and thus prosecutable), but might also have resulted in a more successful prosecution of a doctor who engaged in the same conduct as Quill:

> (3) The patient's attending physician at the time the request is signed shall not be a witness.
> (4) If the patient is a patient in a long term care facility at the time the written request is made, one of the witnesses shall be an individual designated by the facility and having the qualifications specified by the Department of Human Services by rule. [1995 c.3 s.2.02]

This final provision of the written safeguards was an effort by the state to remove assisted suicide from the private sphere (where the potential for abuse is theoretically heightened) to that of the state protected public health sphere.

Section 3 of the Oregon law regarded standards of safeguards that were required to be met before a prescription for lethal medication could lawfully be issued or dispensed. In particular, the law imposed a litany of particularized obligations upon physicians, as follows:

ORS 127.815 s.3.01. Attending physician responsibilities.

> (1) The attending physician shall:

By use of the word "shall" the Oregon statute uses language of mandate, of command, requiring the physician comply with each and every one of the following requirements:

> (a) Make the initial determination of whether a patient has a terminal disease, is capable, and has made the request voluntarily;
> (b) Request that the patient demonstrate Oregon residency pursuant to ORS 127.860;
> (c) To ensure that the patient is making an informed decision, inform the patient of:

(A) His or her medical diagnosis;
(B) His or her prognosis;
(C) The potential risks associated with taking the medication to be prescribed;
(D) The probable result of taking the medication to be prescribed; and
(E) The feasible alternatives, including, but not limited to, comfort care, hospice care and pain control;
(d) Refer the patient to a consulting physician for medical confirmation of the diagnosis, and for a determination that the patient is capable and acting voluntarily;
(e) Refer the patient for counseling if appropriate pursuant to ORS 127.825;
(f) Recommend that the patient notify next of kin;
(g) Counsel the patient about the importance of having another person present when the patient takes the medication prescribed pursuant to ORS 127.800 to 127.897 and of not taking the medication in a public place;
(h) Inform the patient that he or she has an opportunity to rescind the request at any time and in any manner, and offer the patient an opportunity to rescind at the end of the 15 day waiting period pursuant to ORS 127.840;
(i) Verify, immediately prior to writing the prescription for medication under ORS 127.800 to 127.897, that the patient is making an informed decision;
(j) Fulfill the medical record documentation requirements of ORS 127.855;
(k) Ensure that all appropriate steps are carried out in accordance with ORS 127.800 to 127.897 prior to writing a prescription for medication to enable a qualified patient to end his or her life in a humane and dignified manner; and

(l)(A) Dispense medications directly, including ancillary medications intended to facilitate the desired effect to minimize the patient's discomfort, provided the attending physician is registered as a dispensing physician with the Board of Medical Examiners, has a current Drug Enforcement Administration certificate and complies with any applicable administrative rule; or
(B) With the patient's written consent:
(i) Contact a pharmacist and inform the pharmacist of the prescription; and
(ii) Deliver the written prescription personally or by mail to the pharmacist, who will dispense the medications to either the patient, the attending physician or an expressly identified agent of the patient.
(2) Notwithstanding any other provision of law, the attending physician may sign the patient's death certificate. [1995 c.3 s.3.01; 1999 c.423 s.3]

Cumulatively, these provisions imposed many requirements upon doctors. Some of those should be highlighted, and compared with other cases. The Oregon law requires physicians to advise patients about hospice and palliative care (which would effectively have excluded Kevorkian's activities, but embraced Quill's), get a second opinion about the patient's illness, ensure psychological health consistent with patient competency, and repeatedly discuss the consequences of a lethal prescription and the opportunities to rescind (for the patient to change his or her mind) up until the very last second before taking the medication that would result in death. Further, by requiring the patient to give written consent to contact a pharmacist, the law protects against non-voluntary euthanasia as a result of lack of ability.

Similarly, ORS 127.820 s.3.02 imposed requirements of specificity and protocol upon consulting physician confirmation:

> Before a patient is qualified under ORS 127.800 to 127.897, a consulting physician shall examine the patient and his or her relevant medical records and confirm, in writing, the attending physician's diagnosis that the patient is suffering from a terminal disease, and verify that the patient is capable, is acting voluntarily and has made an informed decision. [1995 c.3 s.3.02]

As a juxtaposition, none of Jack Kevorkian's patients would have qualified under this provision, as he did not consult with other doctors. In contrast, Timothy Quill had a variety of external consultations, both oncological and psychological, before he wrote the prescription that resulted in the death of his patient and the unsuccessful prosecution of Quill.

To protect against depressed patients, as well as those with mental health disorders, from unduly obtaining physician-assisted suicide, the Oregon statute also provided for referrals for counseling, by ORS 127.825 s.3.03:

> If in the opinion of the attending physician or the consulting physician a patient may be suffering from a psychiatric or psychological disorder or depression causing impaired judgment, either physician shall refer the patient for counseling. No medication to end a patient's life in a humane and dignified manner shall be prescribed until the person performing the counseling determines that the patient is not suffering from a psychiatric or psychological disorder or depression causing impaired judgment. [1995 c.3 s.3.03; 1999 c.423 s.4]

To compare this to physician prosecutions in the 1990s, Quill had in fact made a psychiatric referral, whereas Kevorkian did not do so in his cases. Thus, for example, had Kevorkian assisted Marjorie Wantz in Oregon (rather than in Michigan), her previous suicide attempts and psychiatric record prior to her seeking him out would have rendered her ineligible under Oregon law (residency and other issues aside):

> ORS 127.830 s.3.04. Informed decision provided that.
>
> No person shall receive a prescription for medication to end his or her life in a humane and dignified manner unless he or she has made an informed decision as defined in ORS 127.800 (7). Immediately prior to writing a prescription for medication under ORS 127.800 to 127.897, the attending physician shall verify that the patient is making an informed decision. [1995 c.3 s.3.04]

By virtue of the definitional constructions of what an "informed consent" was, a number of Kevorkian's patients would have been ineligible under the Oregon law. For example, Marjorie Wantz's diagnosis and prognosis of vulvadenia did not qualify as terminal illnesses that would have feasible alternatives to assisted suicide because medical and/or psychiatric alternatives would have been the appropriate direction. In stark contrast, Quill ensured and elicited the most repeated of informed consents, once approached about assisting in a suicide.

The next Oregon protocol, pertaining to family notification, was designed to encourage (though not mandate) family discussion in the patient's decision-making process. At least in theory, this provision was set to encourage social support, which might mitigate against a patient choosing assisted suicide. This might be viewed as the converse of the protocol prohibiting a family member from being a witness to an assisted-suicide request. In specific, ORS 127.835 s.3.05 provided:

> The attending physician shall recommend that the patient notify the next of kin of his or her request for medication pursuant to ORS 127.800 to 127.897. A patient who declines or is unable to notify next of kin shall not have his or her request denied for that reason. [1995 c.3 s.3.05; 1999 c.423 s.6]

Both written and oral requests are required under the Oregon law, enduringly and repeatedly over a period of no less than two weeks, under ORS 127.840 s.3.06:

> In order to receive a prescription for medication to end his or her life in a humane and dignified manner, a qualified patient shall have made an oral request and a written request, and reiterate the oral request to his or her attending physician no less than fifteen (15) days after making the initial oral request. At the time the qualified patient makes his or her second oral request, the attending physician shall offer the patient an opportunity to rescind the request. [1995 c.3 s.3.06]

The next provision, regarding the right to rescind the request, was seemingly obvious, yet also in essence was a protection against coercion, under ORS 127.845 s.3.07:

> A patient may rescind his or her request at any time and in any manner without regard to his or her mental state. No prescription for medication under ORS 127.800 to 127.897 may be written without the attending physician offering the qualified patient an opportunity to rescind the request. [1995 c.3 s.3.07]

Were Kevorkian to have assisted Hugh Gale in Oregon, under this provision, the moment Gale objected to the gas mask, the proceedings would have had to have ceased and Gale would have had to have been deemed to have rescinded his request for assisted suicide. Perhaps most unusual is that

Chief Prosecuting Attorney Marlinga determined that the case was not prosecutable, although the decision was viewed largely as an informal equivalent to a plea bargain, by the agreement to drop the case in return for the deposition of Gale's widow on a weekend (also highly unusual in timing).

A crucial provision of the Oregon law regarded waiting periods, which applied from the time that the first patient's first to last request was made, and also mandated a minimum two-day waiting period be imposed upon a physician before complying with the patient's final request for assisted suicide and providing the prescription, under ORS 127.850 s.3.08:

> No less than fifteen (15) days shall elapse between the patient's initial oral request and the writing of a prescription under ORS 127.800 to 127.897. No less than 48 hours shall elapse between the patient's written request and the writing of a prescription under ORS 127.800 to 127.897. [1995 c.3 s.3.08]

By this provision, a crucial feature of the assisted-suicide debate was addressed—that a patient may have a fleeting moment or period of despair. Hospice physician Quill's essay showed great sensitivity to this issue. In contrast, in the Kevorkian cases, such waiting periods frequently were not imposed, and in the Youk case (for which Kevorkian was convicted), barely a day had elapsed from the time Kevorkian first met with Tom Youk and Youk's death (which, as euthanasia, would not have been legal under the Oregon statute).

In addition, the Oregon law required full documentation of medical records, and imposed additional protocols to ensure voluntariness on the part of the patient and to ensure against mistaken diagnosis on the part of the doctors under ORS 127.855 s.3.09:

> The following shall be documented or filed in the patient's medical record:
>
> (1) All oral requests by a patient for medication to end his or her life in a humane and dignified manner;
> (2) All written requests by a patient for medication to end his or her life in a humane and dignified manner;
> (3) The attending physician's diagnosis and prognosis, determination that the patient is capable, acting voluntarily and has made an informed decision;
> (4) The consulting physician's diagnosis and prognosis, and verification that the patient is capable, acting voluntarily and has made an informed decision;
> (5) A report of the outcome and determinations made during counseling, if performed;
> (6) The attending physician's offer to the patient to rescind his or her request at the time of the patient's second oral request pursuant to ORS 127.840; and
> (7) A note by the attending physician indicating that all requirements under ORS 127.800 to 127.897 have been met and indicating the steps taken to carry out

the request, including a notation of the medication prescribed. [1995 c.3 s.3.09]

In an effort to control for, and prevent, assisted-suicide tourists, the Oregon statute imposed a residency requirement, with state proof of patient residency to be made under ORS 127.860 s.3.10. These residency proofs went beyond mere indicia, and should be viewed in contrast to the criticism of Keown in his 1997 book, *Euthanasia, Ethics and Public Policy: An Argument against Legalisation*[12]:

Only requests made by Oregon residents under ORS 127.800 to 127.897 shall be granted. Factors demonstrating Oregon residency include but are not limited to:

(1) Possession of an Oregon driver license;
(2) Registration to vote in Oregon;
(3) Evidence that the person owns or leases property in Oregon; or
(4) Filing of an Oregon tax return for the most recent tax year. [1995 c.3 s.3.10; 1999 c.423 s.8]

As an additional control, Oregon imposed state reporting requirements, pursuant to ORS 127.865 s.3.11:

(1)(a) The Health Services shall annually review a sample of records maintained pursuant to ORS 127.800 to 127.897.
(b) The division shall require any health care provider upon dispensing medication pursuant to ORS 127.800 to 127.897 to file a copy of the dispensing record with the division.
(2) The Health Services shall make rules to facilitate the collection of information regarding compliance with ORS 127.800 to 127.897. Except as otherwise required by law, the information collected shall not be a public record and may not be made available for inspection by the public.
(3) The division shall generate and make available to the public an annual statistical report of information collected under subsection (2) of this section. [1995 c.3 s.3.11; 1999 c.423 s.9]

These records both protect the privacy (and hence the dignity) of patients and their families, while providing for governmental supervision and statistical openness. However, in a fair critique, Foley and Hendin, pro-life physicians who are, respectively, a hospice physician and a psychiatrist, level the fair criticism that "nothing in the Oregon law prevents OPHD [the Oregon Public Health Division] from collecting [additional] needed information,"[13] to consider a variety of consequences, such as those in the year after a patient's assisted suicide. They correctly bemoaned this missed opportunity, which, while it might show negative consequences, it could also show positive consequences, such as in the Netherlands, where Swarte et al. found that families of patients who had hastened deaths had the opportunity to say goodbye, which alleviated the grieving process.[14] This might be viewed in stark contrast to the Kevorkian cases, where not only were patient names and personal information disclosed, but tapes were sometimes made available to the media by Kevorkian himself, raising the specter of public invasion of a private domain, with a robbery of patient dignity. What Foley and Hendin would propose is to also seek information from physicians who declined to issue lethal prescriptions, patient psychiatrists (regardless of their findings as to competency), nurses, social workers, and family members. These would yield a trove of information to everyone involved in the debate and who may view Oregon as a test case or laboratory. Foley and Hendin also were critical of the fact that information gathered by the OPHD was, further to protocol, destroyed a year later.

The next provision of the law regarded construction of will and contracts under ORS 127.870 s.3.12:

(1) No provision in a contract, will or other agreement, whether written or oral, to the extent the provision would affect whether a person may make or rescind a request for medication to end his or her life in a humane and dignified manner, shall be valid.
(2) No obligation owing under any currently existing contract shall be conditioned or affected by the making or rescinding of a request, by a person, for medication to end his or her life in a humane and dignified manner. [1995 c.3 s.3.12]

Likewise, the statute provided for insurance (including health insurance, which will be discussed in the epilogue of this chapter) and annuity policies under ORS 127.875 s.3.13:

The sale, procurement, or issuance of any life, health, or accident insurance or annuity policy or the rate charged for any policy shall not be conditioned upon or affected by the making or rescinding of a request, by a person, for medication to end his or her life in a humane and dignified manner. Neither shall a qualified patient's act of ingesting medication to end his or her life in a humane and dignified manner have an effect upon a life, health, or accident insurance or annuity policy. [1995 c.3 s.3.13]

A critical provision, regarding "Construction of the Act," included language prohibiting both active medical euthanasia and mercy killing, under ORS 127.880 s.3.14:

Nothing in ORS 127.800 to 127.897 shall be construed to authorize a physician or any other person to end a patient's life by lethal injection, mercy killing or active euthanasia. Actions taken in accordance with ORS 127.800 to 127.897 shall not, for any purpose, constitute suicide, assisted suicide, mercy killing or homicide, under the law. [1995 c.3 s.3.14]

Although this statute predated Kevorkian's euthanasia of, and prosecution for, the lethal injection of Tom Youk, this provision served to put strict limitations on what conduct would and would not be deemed to be patient death with dignity. Had Tom Youk died by similar circumstances in Oregon in 1998, the doctor would have been fully prosecuted under the law. Thus appears the solution to one of the seeming paradoxes of the Oregon assisted-suicide law—by making assisted suicide legal and creating protocols under which lawful assisted suicide could be practiced, the law also created safeguards protecting against that which could not be practiced. In other words, the statute also provided for de jure criminalization of euthanasia and mercy killing.

In addition, the Oregon law provided for immunity from civil and criminal liability, as well as from professional disciplinary action for those physicians who complied with patient requests in good faith, in Section 4 of ORS 127.885 s.4.01. In specific, except as provided in ORS 127.890:

> (1) No person shall be subject to civil or criminal liability or professional disciplinary action for participating in good faith compliance with ORS 127.800 to 127.897. This includes being present when a qualified patient takes the prescribed medication to end his or her life in a humane and dignified manner.

As a side comment, this provision specified that mere presence is not enough to give rise to prosecution or to civil liability. Another way of saying this is that the patient must engage in the final act of taking the medication, which the subsection requires be duly prescribed (the breach of which implicates drug delivery laws, as well) and that the patient must be "qualified" and fully eligible as discussed above. Thus, if a patient does not meet the qualifications, a physician may be held liable under the Oregon law for providing the means and opportunity to end the patient's life, even if the patient engaged in the final act.

> (2) No professional organization or association, or health care provider, may subject a person to censure, discipline, suspension, loss of license, loss of privileges, loss of membership or other penalty for participating or refusing to participate in good faith compliance with ORS 127.800 to 127.897.
> (3) No request by a patient for or provision by an attending physician of medication in good faith compliance with the provisions of ORS 127.800 to 127.897 shall constitute neglect for any purpose of law or provide the sole basis for the appointment of a guardian or conservator.
> (4) No health care provider shall be under any duty, whether by contract, by statute or by any other legal requirement to participate in the provision to a qualified patient of medication to end his or her life in a humane and dignified manner. If a health care provider is unable or unwilling to carry out a patient's request under ORS 127.800 to 127.897, and the patient transfers his or her

care to a new health care provider, the prior health care provider shall transfer, upon request, a copy of the patient's relevant medical records to the new health care provider.

In effect, this subsection was a "conscientious objector" clause, so that those who object to assisted suicide for reasons of theological or ethical or moral belief are not required to participate, similarly to abortion provisions.

(5)(a) Notwithstanding any other provision of law, a health care provider may prohibit another health care provider from participating in ORS 127.800 to 127.897 on the premises of the prohibiting provider if the prohibiting provider has notified the health care provider of the prohibiting provider's policy regarding participating in ORS 127.800 to 127.897. Nothing in this paragraph prevents a health care provider from providing health care services to a patient that do not constitute participation in ORS 127.800 to 127.897.

This was likewise an opt-out, for a hospital, nursing home or other provider.

(b) Notwithstanding the provisions of subsections (1) to (4) of this section, a health care provider may subject another health care provider to the sanctions stated in this paragraph if the sanctioning health care provider has notified the sanctioned provider prior to participation in ORS 127.800 to 127.897 that it prohibits participation in ORS 127.800 to 127.897:

This provision in essence said that should there be a difference of opinion or philosophy as to whether or not a patient may receive assistance in suicide, a health care provider (hospital, nursing home, etc.) that does not permit assisted suicide can discipline a physician or other employee who nonetheless participates in an assisted suicide. This removed from debate the question of whether an institution which has a pro-life or non-assisted-suicide policy, by resolving the issue in favor of life. Sanctions may be severe, yet those institutions and individuals who opt not to participate in death with dignity, or assisted suicide, also have enumerated opportunities to give referrals, as follows:

(A) Loss of privileges, loss of membership or other sanction provided pursuant to the medical staff bylaws, policies and procedures of the sanctioning health care provider if the sanctioned provider is a member of the sanctioning provider's medical staff and participates in ORS 127.800 to 127.897 while on the health care facility premises, as defined in ORS 442.015, of the sanctioning health care provider, but not including the private medical office of a physician or other provider;

(B) Termination of lease or other property contract or other nonmonetary remedies provided by lease contract, not including loss or restriction of medical staff privileges or exclusion from a provider panel, if the sanctioned provider participates in ORS 127.800 to 127.897 while on the premises of the sanctioning health care provider or on property that is owned by or under the direct control of the sanctioning health care provider; or

(C) Termination of contract or other nonmonetary remedies provided by contract if the sanctioned provider participates in ORS 127.800 to 127.897 while acting in the course and scope of the sanctioned provider's capacity as an employee or independent contractor of the sanctioning health care provider. Nothing in this subparagraph shall be construed to prevent:

(i) A health care provider from participating in ORS 127.800 to 127.897 while acting outside the course and scope of the provider's capacity as an employee or independent contractor; or

(ii) A patient from contracting with his or her attending physician and consulting physician to act outside the course and scope of the provider's capacity as an employee or independent contractor of the sanctioning health care provider.

(c) A health care provider that imposes sanctions pursuant to paragraph (b) of this subsection must follow all due process and other procedures the sanctioning health care provider may have that are related to the imposition of sanctions on another health care provider.

(d) For purposes of this subsection:

(A) "Notify" means a separate statement in writing to the health care provider specifically informing the health care provider prior to the provider's participation in ORS 127.800 to 127.897 of the sanctioning health care provider's policy about participation in activities covered by ORS 127.800 to 127.897.

(B) "Participate in ORS 127.800 to 127.897" means to perform the duties of an attending physician pursuant to ORS 127.815, the consulting physician function pursuant to ORS 127.820 or the counseling function pursuant to ORS 127.825. "Participate in ORS 127.800 to 127.897" does not include:

(i) Making an initial determination that a patient has a terminal disease and informing the patient of the medical prognosis;

(ii) Providing information about the Oregon Death with Dignity Act to a patient upon the request of the patient;

(iii) Providing a patient, upon the request of the patient, with a referral to another physician; or

(iv) A patient contracting with his or her attending physician and consulting physician to act outside of the course and scope of the provider's capacity as an employee or independent contractor of the sanctioning health care provider.

(6) Suspension or termination of staff membership or privileges under subsection (5) of this section is not reportable under ORS 441.820. Action taken pursuant to ORS 127.810, 127.815, 127.820 or 127.825 shall not be the sole basis for a report of unprofessional or dishonorable conduct under ORS 677.415 (2) or (3).

(7) No provision of ORS 127.800 to 127.897 shall be construed to allow a lower standard of care for patients in the community where the patient is treated or a similar community. [1995 c.3 s.4.01; 1999 c.423 s.10]

This last provision in effect removed from debate the possibility of engaging in neglectful conduct so as to promote death of a patient to whom assisted suicide cannot be granted.

Further, while the Oregon Death with Dignity Act is viewed, correctly, as allowing for assisted suicide and providing for protocols for lawful physician suicide, the statute also enumerated criminal penalties for those who engage in conduct not within the legal protocols and procedures. ORS 127.890 s.4.02 enumerated these criminal and potential civil liabilities, which placed noncompliant physician behavior in the most severe class of felonies.

(1) A person who without authorization of the patient willfully alters or forges a request for medication or conceals or destroys a rescission of that request with the intent or effect of causing the patient's death shall be guilty of a Class A felony.
(2) A person who coerces or exerts undue influence on a patient to request medication for the purpose of ending the patient's life, or to destroy a rescission of such a request, shall be guilty of a Class A felony.
(3) Nothing in ORS 127.800 to 127.897 limits further liability for civil damages resulting from other negligent conduct or intentional misconduct by any person.
(4) The penalties in ORS 127.800 to 127.897 do not preclude criminal penalties applicable under other law for conduct which is inconsistent with the provisions of ORS 127.800 to 127.897. [1995 c.3 s.4.02]

In addition to the other protocols set for physicians and others to show compliance with the enumerated protocols, the Oregon Death with Dignity Act provided for a standard "form of the request" to be completed by patients who requested assisted suicide. This form, in Section 6, required patients to give information prior to receiving a prescription, as stated in ORS 127.897 s.601:

A request for a medication as authorized by ORS 127.800 to 127.897 shall be in substantially the following form:

REQUEST FOR MEDICATION TO END MY LIFE IN A HUMANE AND DIGNIFIED MANNER

I, __________________, am an adult of sound mind.

I am suffering from ________, which my attending physician has determined is a terminal disease and which has been medically confirmed by a consulting physician.

I have been fully informed of my diagnosis, prognosis, the nature of medication to be prescribed and potential associated risks, the expected result, and the feasible alternatives, including comfort care, hospice care and pain control.
I request that my attending physician prescribe medication that will end my life in a humane and dignified manner.
INITIAL ONE:
_____ I have informed my family of my decision and taken their opinions into consideration.
_____ I have decided not to inform my family of my decision.
_____ I have no family to inform of my decision.
I understand that I have the right to rescind this request at any time.
I understand the full import of this request and I expect to die when I take the medication to be prescribed. I further understand that although most deaths occur within three hours, my death may take longer and my physician has counseled me about this possibility.
I make this request voluntarily and without reservation, and I accept full moral responsibility for my actions.
Signed: ___________
Dated: ___________
DECLARATION OF WITNESSES
We declare that the person signing this request:
(a) Is personally known to us or has provided proof of identity;
(b) Signed this request in our presence;
(c) Appears to be of sound mind and not under duress, fraud or undue influence;
(d) Is not a patient for whom either of us is attending physician.
__________ Witness 1/Date
__________ Witness 2/Date
NOTE: One witness shall not be a relative (by blood, marriage or adoption) of the person signing this request, shall not be entitled to any portion of the person's estate upon death and shall not own, operate or be employed at a health care facility where the person is a patient or resident. If the patient is an inpatient at a health care facility, one of the witnesses shall be an individual designated by the facility.
[1995 c.3 s.6.01; 1999 c.423 s.11]

Last, ORS 127.995 provided for penalties to be applied in cases where written instruments and/or evidence was altered, regarding felony penalties in cases of withdrawal of nutrition and hydration (usually an issue in PVS cases, rather than assisted suicide) and both misdemeanor and felony penalties in cases of health care decisions (which included assisted suicide in its ambit). These provisions stated:

(1) It shall be a Class A felony for a person without authorization of the principal to willfully alter, forge, conceal or destroy an instrument, the reinstatement or revocation of an instrument or any other evidence or document reflecting the

principal's desires and interests, with the intent and effect of causing a withholding or withdrawal of life-sustaining procedures or of artificially administered nutrition and hydration which hastens the death of the principal.

(2) Except as provided in subsection (1) of this section, it shall be a Class A misdemeanor for a person without authorization of the principal to willfully alter, forge, conceal or destroy an instrument, the reinstatement or revocation of an instrument, or any other evidence or document reflecting the principal's desires and interests with the intent or effect of affecting a health care decision. [Formerly 127.585]

CONCLUSION: CONSTRUCTING AN OREGONIAN EPILOGUE AS A JUXTAPOSITION OF THE OREGON EXPERIENCE

On June 3, 2011, Jack "Dr. Death" Kevorkian died of natural causes relating to liver failure and pneumonia, in a hospital. Notwithstanding the natural death of the world's most famous assisted-suicide and euthanasia advocate, the assisted-suicide debate was heating up during that very period. HBO Documentary Films' *How to Die in Oregon* debuted for general release less than a week before Kevorkian himself died. The documentary offered a (largely pro-choice) contribution to this debate. Peter D. Richardson spent approximately four years (starting in 2007, approximately the same time that Jack Kevorkian was released on parole in Michigan, under stringent terms including not advising about or assisting in suicide or euthanasia) and $750,000 working on this film. *How to Die in Oregon* was extremely difficult to watch, regardless of whether viewers consider themselves pro-choice or pro-life or a little bit of both or a lot of neither. Perhaps that is a testament to the excellence of this film, the winner of the 2011 Sundance Grand Jury Prize for Documentary.

The 108-minute long HBO film was intimate, perhaps too much so, depicting lawful assisted suicides in Oregon at both the beginning and the end of the film. Opening with the suicide of Roger Sagner, the 343rd person to make use of the Oregon Death with Dignity Law, viewers were reminded of *The Sea Inside*, Alejandro Amenabar's 2004 fictional depiction of the life and death of quadriplegic Ramon Sampedro. After Spanish courts denied Sampedro's requests that those who helped him to die not be prosecuted, he nevertheless ended his life with the help of "many hands," who variously mixed a lethal potion, gave it to him to drink through a straw, and spent the last moments of his life with him.

Unlike the fictionalized film, *How to Die in Oregon* showed the real Sagner and the actual death. It also showed a colloquy with a volunteer from Compassion & Choices (of which Barbara Coombs Lee, one of the original authors and petitioners of Measure 16, was the president), who mixed the lethal cocktail for him, after asking him whether he had changed his mind and ensuring that he knew what the medication will do, a protocol employed by volunteers in such cases. Sagner spent his last moments with his friends, who stayed with him while he drank the cocktail (which he proclaimed "tastes like poison") and until after he died. Perhaps this scene provided a partial answer to the reporting issues Foley and Hendin raised.

This opening scene demonstrated a number of differences between Kevorkian's practice (and Quill's experience, for that matter) and how the Oregon law was constructed and implemented. First, while the prescription was written by the doctor, the Compassion & Choices volunteer was not a doctor; hence, a depiction of physician-assisted suicide by prescription, or by provision of the means, which the patient then deployed. Second, Sanger was surrounded by loved ones, rather than dying alone (as did Tom Youk, for whose euthanasia murder Kevorkian was ultimately convicted). Third, the patient decided when and how the drugs would be administered, and controlled the final acts; this absolute requirement of the Oregon law was later depicted in the HBO film as a reason one man was unable to avail himself of physician-assisted suicide.

The HBO documentary is exactly six times as long, or 600 percent longer, than the "Death By Doctor" segment that led to Kevorkian's final prosecution and conviction. As discussed in Chapter 5, the final indictment and trial was sought by a previously reluctant prosecutor, who in fact had dismissed assisted-suicide charges brought by his predecessor, and who only initiated the proceedings that resulted in the 1999 trial because of the public outrage that resulted from CBS broadcast of the Youk euthanasia, coupled by the Kevorkian narration and interview. Moreover, the HBO film showed not one (as with the Kevorkian "Death by Doctor" segment), but two actual (not fictionalized) hastened deaths. Thus, a question emerges as to whether the historical perspective of whether assisted death has emerged from the shadows and gone beyond the intended private sphere that the Oregon legislation contemplated, and into a public one. In any event, the headline-inducing shock value of Kevorkian's Youk euthanasia was not matched by the HBO film, which was more graphic in some respects.

How to Die in Oregon offered an epilogue, but also a prologue, as to introducing the Washington assisted-suicide ballot movement and debate. Interwoven with the stories of Oregonians who wanted to make use of the Oregon Death with Dignity law was the story of Nancy Niedzielski, a Seattle woman who moved the (successful) Washington I-1000 campaign forward. While the Washington debate will be discussed in the next chapter, regarding post-millennium legislation and litigation, her story fits into this Oregon epilogue, too. Niedzielski's husband was not eligible under the Oregon law, because he could not establish legal residency between the time he was diagnosed with brain and spinal cord cancer and the time of his death the next year. In contrast to this story is that of Randy Stroup, who went to the press after his insurance company wrote him, saying that it would cover a prescription for medical aid in dying, but not cover the cost of additional therapy (the health insurance company reversed itself in the wake of the press coverage); this was in clear violation of the statutory provisions relating to health insurance.[15] Yet another perspective is given by the story of 84-year old Ray Carny, who ultimately died in the hospital, physically unable to give himself the lethal cocktail, one of the requirements of the law.[16]

Much of the film focuses on Cody Curtis, who valiantly battled liver cancer, before availing herself of the same process at the end of the film—called Death with Dignity (after the name of the legislation), and pointedly not called assisted suicide.

While there was commentary by both Derek Humphry, and his "how-to" book and DVD/video *Final Exit*, and pro-life advocates, this film belonged to the patients, and was told by them and from their perspective, a unique contribution. Family members are present, as are some of the physicians, but the doctors are not the stars (as was Kevorkian), but rather supporting actors. If Kevorkian's tragic flaw was attempting to be his own director and producer, then HBO's film was flawless, although still tragic in the depicted deaths, and not for the faint of heart.

Comparing the developing history of the Oregon experience (as described by the patients in the HBO film), and the completed Michigan assisted-suicide and euthanasia history (as defined by the death of Kevorkian, not of any of his patients or clients), one matter of debate seems resolved. The patients in Oregon in Richardson's film were surrounded by family and friends when they were assisted (by non-physician volunteers) in their suicides, whereas Kevorkian's patients were often alone and bereft of the presence and comfort of their loved ones when they died. That was not always the case, as Merian Frederick had both family and clergy present when she was assisted by Kevorkian; however, family became less visible at the time of death, partly so as to avoid accessorial prosecution. This said, it is also debatable whether Kevorkian was also seeking to have greater control over the death scenarios, as he argued both to Mike Wallace in the "Death by Doctor" segment and (unsuccessfully) to the jury at trial.

Ultimately, the Oregon experience is ongoing in nature, but there were immediate differences between that of Oregon and Michigan. During the first year, of the 23 people who received prescriptions (and the 15 who used them, unlike the 6 who died of underlying illness and the 2 who had not yet used them), the median age was 69, half were men and half women (a great difference from the whopping 72 percent of Kevorkian patients who were women),[17] all were Caucasian, and some 13 had cancer (and, by requirement of statute, had months or less left to live).[18] The patients were reportedly "similar with regard to sex, race, urban or rural residence, level of education, health insurance coverage and hospice enrollment [and] no case patients . . . expressed concern about the financial impact of their illness"[19] One social attribute that perhaps went underexplored was that "the case patients were more likely than the control patients to have never married."[20] This last was in keeping with sociological statistics about traditional suicide developed by sociologist Emile Durkheim over a century earlier.[21] While only one of the first-year patients expressed concern about inadequate pain control, most were likely to be concerned about loss of autonomy due to illness and loss of control of bodily functions.[22]

The second-year experience was similar, with a proportional decrease in cases of cancer (17 of the 27 patients in 1999) and a proportional increase of those with ALS (four) and one of chronic obstructive pulmonary disease.[23] Perhaps the most encouraging statistic was that of the 27 patients who died after taking lethal medications, all had health insurance, and some 21 were in hospice care.[24] This seems to suggest that Oregonians were approaching assisted suicide as a last resort, only after exhausting hospice and palliative care options. Indeed, the general numeric trends pertaining to age, education, marital status, and insurance were stable; but over time, the 1997–2000 proportion of patients who were enrolled in hospice was 87 percent.[25] One might credit this to the three-year delay in implementation of the law, while *Lee v. Oregon* was pending, and during which time the Oregon Health Sciences University Center for Ethics in Health established a multidisciplinary task force "to promote excellent care of the dying and address both ethical and clinical issues inherent in implementing the DWDA [*sic*], . . . published a guidebook . . . for health care professionals and institutions to use as a resource for examining the implications" of implementing the Oregon Death with Dignity Act.[26]

The Oregon Public Health Division report on the statistics for the Oregon Death with Dignity Act use during 2011 showed that the earlier trends continued. While 114 prescriptions were written, only 97 were used.[27] This suggests an ongoing pattern of comfort in simply having the prescriptions, without a necessity for using them. As further support for this proposition is that nine of the patients who ingested the medication during 2011 had had prescriptions written in previous years, and some 25 of the patients who requested the prescriptions ultimately died of their underlying illnesses.[28] Of the 71 people who died in 2011 using prescriptions under the

Oregon Death with Dignity Act, some 69 percent were over the age of 65 (with a median age of 70), 95.6 percent were white, 58.5 percent had a college degree (or higher), and some 83.4 percent had cancer.[29] Perhaps most important, 96.7 percent of the patients who died by use of the Oregon Death with Dignity Act were enrolled in hospice at the time they took the lethal dose.[30] No referrals were made to the Oregon Medical Board for physician failure to comply with state protocols.

The Oregon experience has, thus far, proven to be unique in the passage of legislation allowing for physician-assisted suicide, having its law upheld and having a stable practice that did not find itself on Kamisar's thin edge of a wedge. That may make it a positive example for those states seeking to permit assisted suicide, although it will never answer the thorny question of preservation and sanctity of life fully for those who have a religious basis for opposing the practice, as St. John-Stevas did. It would, however, be held to be an exemplar of the positive construction of the debate set forth by Glanville Williams, half a century earlier.

As a postscript to the Oregon discussion, Dr. Peter Goodwin, who was one of the early champions of the Oregon law, at age 83, himself died after using a lethal prescription under the Oregon law, after battling a rare brain disorder diagnosed in 2006.[31] He was reported to have been surrounded by family members at the time of his March 2012 self-administered death. Dr. Goodwin's death is in stark contrast to that of Dr. Jack Kevorkian, whose June 2011 death, by natural causes, took place in a hospital.

NOTES

1. Hillyard and Dombrink, *op cit.*, p. 73.
2. Stutsman, Eli, "Letter to Peter Norton," Regarding the Oregon Death with Dignity Act, June 2, 1994 (cited in Hillyard and Dombrink, *op. cit.*, 73–74).
3. O'Keefe, Mark, "Founding Father: Derek Humphry Began the Assisted Suicide Movement, but his Views May Be Too Extreme for Measure 16 Strategists," *Oregonian*, November 2, 1994, C1.
4. Ayres, B. Drummond, "The 1997 Elections Referendums: Oregon Stays With Its Law on Suicides," *New York Times*, November 5, 1997.
5. Hillyard and Dombrink, *op. cit.*, 70.
6. *Lee v. Oregon*, 107 F.3d 1382 (3d Cir. 1997).
7. Garrow, David J., "The Oregon Trail," *New York Times*, November 6, 1997.
8. *Gonzales v. Oregon*, 546 U.S. 243 (2006).
9. 21 U.S.C. Sections 801 *et seq.*
10. 546 U.S. 243 (2006).
11. Biskupic, Joan, "Opinion More or Less Rebukes Power Grab: Justice Cites Politics Behind Directive," *USA Today*, January 18, 2006, 5A.

12. Keown, J., *Euthanasia, Ethics and Public Policy: An Argument against Legalisation* (Cambridge: Cambridge University Press, 1997), 168.

13. Hendin Herbert, and Kathleen Foley, *Michigan Law Review* 106 (2008): 1613–39.

14. Swarte, N. V., M. I. van der Lee, J. G. van der Boom, J. van den Bout, and A. P. Heintz, "Effects of Euthanasia on the Bereaved Family and Friends: A Cross-Sectional Study," *British Medical Journal* 327, no. 7408 (July 26, 2003): 189.

15. ORS 127.875 s. 3.12 [1995 c.3 s.3.13].

16. See ORS 127.805 s.2.01.

17. Dragovic et al., *op. cit.*

18. Chin, Arthur E., Katrina Hedberg, Grant K. Higginson, and David W. Fleming, "Legalized Physician-Assisted Suicide in Oregon—the First Year's Experience," *New England Journal of Medicine* 340 (February 18, 1999): 577–83.

19. *Id.*

20. *Id.*

21. Durkheim, Emile, *Suicide: A Study in Sociology*, translated from French by John A. Spaulding and George Simpson. (New York: The Free Press, 1966/1952).

22. Chin et al., *op. cit.*

23. Sullivan, Amy D., Katrina Hedberg, and David W. Fleming, "Legalized Physician-Assisted Suicide in Oregon—the Second Year," *New England Journal of Medicine* 342 (February 24, 2000): 598–604.

24. *Id.*

25. Volker, Deborah L., "The Oregon Experience with Assisted Suicide," *Journal of Nursing Law* 11, no. 3: 152–62.

26. *Id.*

27. http://public.health.oregon.gov/ProviderPartnerResources/EvaluationResearch/DeathwithDignityAct/Documents/year14.pdf, visited March 18, 2012.

28. *Id.*

29. *Id.*

30. *Id.*

31. AP Staff, "Oregon Physician behind Death with Dignity Law Dies," *Oregonian*, March 12, 2012.

8

The 2000s: Post-Millennium Case Studies of State Reactions to Euthanasia and Assisted Suicide

INTRODUCTION

The landscape of the American medical euthanasia and physician-assisted suicide debate was topographically rich in the decade after the millennium. These ranged from coast to coast, and variously involved different medical activities. In 2002, a woman engaged in a double mercy killing of two sons in a Georgia medical setting, yet the shooting death was treated unlike a typical murder or violent crime. Then, as if taking up the 1997 invitation of the U.S. Supreme Court in the *Glucksberg* and *Quill* cases, for states to individually deal with these issues, and taking a lead from the 2006 *Gonzales* case affirming the Oregon Death with Dignity Act for reasons of a narrow procedural nature, a clutch of states had their own brushes with dealings regarding doctors and their conduct. In Louisiana, a doctor was prosecuted (but ultimately not indicted for) medical euthanasia in 2005, after Hurricane Katrina and its unique necessities. Sprawling over the first half of the decade, a family feud regarding (in part) whether the husband or parents of Terri Schiavo should have decision making authority was couched in terms of how to handle a Florida woman in a persistent vegetative state. In 2009, a statute in Washington State was modeled on the Oregon law (and, like Oregon, further to a ballot initiative to legalize assisted suicide). Also in 2009, a successful civil suit brought by a dying man in Montana, who died hours before the first court decision affirming his right under state law to have assisted suicide, and to have any doctor involved be absolved of criminal liability for assisting.

Treated in series, they have in common that by relying upon state law, the cases fall outside the federal province, and have medical foci.

THE GEORGIA MERCY KILLINGS BY, AND ASSISTED-SUICIDE PLEA BARGAIN, OF CAROL CARR (2002)

A mercy killing (or, more accurately, a simultaneous pair of mercy killings) by a parent might not initially seem to be the sort of case to fit into a project focused upon the conduct of physicians. However, Carol Carr shot her two sons, who had been debilitated by Huntington's disease. Andy Byron Scott, 41, and Michael Randy Scott, 42, were unable to walk, feed themselves, or engage in lucid thought.

Andy and Michael Scott, like their father (Carr's first husband) were afflicted with Huntington's disease, a dominant genetic progressive degenerative illness with a midlife onset, which robs its victims of physical ability and mental lucidity over a typical period of 10–20 years, and which does not generally have a terminal illness definition by anticipated date or time of death in the six-months-to-one-year range most frequently considered in assisted-suicide cases. Because the case went untried (and, without conviction after trial, without a transcript of evidence adduced during the non-trial) reports of this case remain based upon (primarily local) press. At the time of the brothers' deaths caused by their mother's gun shooting, they lay side by side in beds at a nursing home in a small town of 20,000 in Bible-Belt Georgia, when their mother walked into the room and shot them.[1] This suggests late-stage (if not end-stage) illness, including lack of physical mobility and lack of mental capacity (even assuming a state where assisted suicide was lawful, further to voluntary consent of an adult capable of so engaging in).

There was no question that the brothers died as a result of the shootings and not of any intervening cause of death. Such a potential *novus actus intervenus*—even if viewed generally as a legal fiction—has been used since the New Hampshire trial of Hermann Sander in the 1950s, as in the Kevorkian/Youk trial in 1999, when there was a possible (albeit unsuccessful) issue that Tom Youk had died of his end-stage ALS, or Nigel Cox's 1992 trial, where the possibility that Lillian Boyes had died of her underlying end-stage rheumatoid arthritis, rather than due to the potassium chloride that Cox unquestionably injected, mitigating against a murder charge and resulting in one for attempted murder. In Carr's case, there was no room for consideration of causation; neither was there any question that Carol Carr fully intended to end their lives by shooting the gun at each son, in succession. Thus, the two typical elements of homicide, causation and intention, were met as to Carr.

Even assuming for the sake of argument that Georgia at the time had a law providing for physician-assisted suicide, this case was outside the ambit for several reasons. First, there was the obvious issue that the patients were not the persons who engaged in the final act—they took no prescription, as would have been allowed under Oregon's Death with Dignity Act. Second, and perhaps less obviously played out in the press, the sons did not have the capacity to consent to their deaths, even assuming that the typical prohibition and policy forbidding consent to one's own homicide was ignored; indeed, "the degree of participation"[2] was further than "merely aiding in someone else's active decision"[3] given that Carr herself took the decision to kill her two sons. Thus, as compassionate as her motive may have been (and which went unquestioned by her townspeople), her action was not altogether different than that contemplated by Burleigh in his 1994 work on the Nazi euthanasia program, and who should get the so-called benefit.

There is also a jump as to the fact that Carr engaged in a double homicide; indeed, Jack Kevorkian lost his medical license, as discussed in Chapter 5, after he engaged in the 1991 double assisted suicide of Marjorie Wantz and Sherry Miller, conduct for which he was also indicted for murder (although acquitted in 1996). Legislation enacted in other states as a response to the issues that arose from Kevorkian's activities included a statute in Georgia, regarding "offering to assist in the commission of a suicide,"[4] which was used for the first time to allow for the plea bargaining of Carr. A Spalding County grand jury reportedly indicted Carr in August 2002 on two charges of felony murder and two charges of malice murder, even as "the surrounding community has reached out to the family, which has received hundreds of supportive letters."[5] This presented a further irony, anticipating the findings of Swarte et al., that family members of those who had euthanasia deaths in the Netherlands exhibited less complicated grief if they had an opportunity to say goodbye (although the study contemplated medical euthanasia deaths versus non-euthanasia deaths, and not family mercy killings).

Nonetheless, Carr was permitted to enter a plea of guilty to two charges of assisting in the commission of a suicide, and served two years of the five-year sentence to which she was originally sentenced. Further, she was permitted to serve a five-year sentence of probation on the second count, on condition that she not live with her 40-year-old son James, also afflicted with the early stages of Huntington's disease. This creative prosecutorial use of plea bargaining as a compassionate legal tool for a woman who was neither physician nor assisting in the fatal act harkens back to Kamisar's commentary that the law in action was as malleable as the law on the books was absolute, and as an example of the way forward as criminal law and the criminal justice system seek to deal with a newly emerging issue of family-assisted suicide and prosecutorial responses thereto.

It is something of an irony that the prosecutorial compassion in the Carr case was notwithstanding numerous aberrations in the criminal law, and indeed in recent Georgia law. The use of plea bargaining for assisted suicide, in a case of shooting murder (double at that), by a nonmedical person in a medical facility, of two sons who were rendered incompetent by the level of their advanced state of Huntington's disease, all demonstrate a post-millennial perspective of a what historically would have been either tried (or pleaded to) as mercy killing, i.e., homicide, with its lengthy sentencing (in a death penalty state).

THE UNTRIED POST-KATRINA EUTHANASIAS BY DR. ANNA POU

The political and criminal justice applications of the law regarding euthanasia took an unprecedented turn in 2005 post-Katrina New Orleans. In specific, the natural disaster provided an unlikely perfect storm of elements, which led to Dr. Anna Pou ordering and/or giving lethal injections to at least four (and possibly as many as 17) dying patients at Uptown's Memorial Hospital, while they awaited a long-delayed post-Hurricane rescue. Nevertheless, Pou was not indicted by a grand jury, and an article by physician turned investigative reporter Sherri Fink, written for the *New York Times Sunday Magazine* about how the disaster precipitated a perfect storm that resulted in medical euthanasia by Pou and by nurses at Memorial Hospital, won the Pulitzer Prize for investigative reporting several years later, in 2009.[6]

The doctor and nurses who were investigated in connection with the four (definite) euthanasia deaths (out of some 45 at Memorial Hospital during the aftermath of the hurricane) had been under severe stress in post-Katrina New Orleans, with over 100 degrees and little (if any) sleep for several days. While arrested and investigated, Pou, a Catholic, maintained her position that her job was to "help" the patients "through their pain."[7] The injections contained morphine (commonly viewed as a drug with the double effect of suppressing respiration) and/or the sedative midazolam.

In a sense Pou was arguing a case of necessity (or force majeure), which as a matter of Anglo-American criminal law, is generally considered to be an excuse to be used in mitigation of sentencing (after conviction) rather than as a justification (which would result in an acquittal at trial). Here, the act of God of the hurricane was compounded by the lack of rescue after several days, which had cascading consequences. The killings would be considered to be within the "textbook example of the wrongful killing of an innocent person that might be excused on grounds of necessity."[8]

In essence, Pou was a non-publicity-seeking version of New York's Quill—someone who engaged in unlawful conduct, who was walked out by the grand jury despite the facts. While Quill's conduct was that of assisting in suicide, and Pou's was unquestionably euthanasia (of a non-voluntary nature), the analogy of the grand jury's decision-making process appears to have been similar. Both appeared to have been focused upon extraordinary situations that had emerged before them (although Quill was investigated and initially prosecuted because of his *New England Journal of Medicine* article, in contrast to the Pou investigation).

Pou's euthanasia of hospital patients begs the question of juxtaposition to the Kevorkian cases, most especially the 1999 Kevorkian prosecution, for which he was convicted and incarcerated for the murder of Tom Youk. Pakes observed that "[p]rosecution is about filtering out cases that should not go to court."[9] While the Louisiana prosecutor sought second-degree murder charges, upon facts that seemed destined to result in sufficient proof of the elements, the grand jury effectively engaged in pre-trial grand jury nullification, a more unusual filtering. Thus once again, Kamisar's malleable law in action was seen, as a result of the common sense of a local population that had had to experience firsthand the many deprivations and emergencies that Katrina and her aftermath had to offer the Gulf states. The March 2007 grand jury's refusal to indict Pou, the first female doctor to be accused of euthanasia, may have had as much to do with the politics of natural disasters as with euthanasia. Whereas ordinarily a grand jury may have focused upon the question Fink raised as to whether Pou performed comfort care (palliative care with a secondary double effect) or mercy killing, the fact that Pou was a physician removed the question of mercy killing from the minds of the jurors, who likely decided medical murder or no, necessity or no, a stand against Katrina and how New Orleans was neglected or no.

Tellingly, Fink found that in the years after Pou's brief prosecution, she has spent a good deal of her time seeking legal change, and that in particular, she "helped write and pass three laws in Louisiana that offer immunity to health care professionals from most civil lawsuits—though not in cases of willful misconduct—for their work in future disasters, from hurricanes to terrorist attacks to pandemic influenza."[10] While her efforts in these laws "also encourage prosecutors to await the findings of a medical panel before deciding whether to prosecute medical officials,"[11] an intriguing comparison is to be made between Pou's case and the mid-1990s Kevorkian prosecutions, particularly that in Wayne County in 1994, when Chief Prosecuting Attorney John O'Hair effectively hailed Kevorkian for service in bringing an issue to the attention of the public, and perhaps limit his efforts to advocacy as to euthanasia (ironically, O'Hair was also an advocate of what he referred to as assisted suicide by lethal injection or, more accurately, medical euthanasia). That Pou went on to this softer approach, with its focus upon emergency situations (rather than voluntary euthanasia or assisted suicide generally of a patient with a terminal illness) may place her in a higher credibility category (and certainly does so when compared to a Jack Kevorkian), particularly when taking her euthanasia experience into account, notwithstanding the fact that her advocacy efforts are not general as to medical euthanasia or assisted suicide. While this has not generally been discussed in the euthanasia or physician-assisted suicide debate, the narrow parameters of this exclusion from criminal liability serve as furthering the debate in a substantive way, capable of repetition, yet evading general review, but for this advocacy.

THE 2005 CASE OF TERRI SCHIAVO (FLORIDA)

Terri Schiavo "may have had the most public death of any private person in history."[12] Indeed, Schiavo was to PVS what Jack Kevorkian was to assisted suicide: a poster child. However, whereas in each of Kevorkian's 130+ cases, he required family unity and family consent, the Schiavo case represented dispute and disagreement between Terri Schiavo's husband Michael and her parents, the Schindlers. In addition, the Schiavo case is neither a case of voluntary medical euthanasia nor a case of assisted suicide. However, the case may be used to point to some of the issues that emanate from such cases, even in the post-millennium era.

Schiavo spent most of her teenage years as extremely overweight, then transformed herself from a 250-pound woman to one who was slim, meeting and marrying Michael Schiavo along the way. After they married, she moved to Florida, and, according to her guardian *ad litem*, Jay Wolfson, "she continued to lose weight aggressively, until six years after their marriage, she was only 110 pounds."[13]

On February 5, 1990, Schiavo sustained a cardiac arrest and also suffered anoxia, with her brain being deprived of oxygen for between 10 and 12 minutes, or "nearly twice as long as is generally necessary to cause profound, irreversible brain damage."[14] Over the next few years, Michael Schiavo sought to have extraordinary medical care, including transporting Terri to California for experimental treatment that went without success. During this time, husband Michael and Terri's parents, the Schindlers, remained friends. Also during this time, there was a successful medical malpractice suit against Terri's fertility specialist, who sustained a jury judgment against him in civil court for malpractice to the effect that had Terri's potential eating disorders been diagnosed and treated, she might not have had the collapse that led to her persistent vegetative state. From this, Michael was awarded $300,000 (for loss of consortium) and another $700,000 was placed in trust for Terri's future care and maintenance, all with "no hope of rehabilitation . . . a family tragedy."[15]

As played out in the media, Michael Schiavo, who had been Terri's guardian, sought in 1998 to remove her feeding tube, by which she received artificial nutrition and hydration; to this parents Robert and Mary Schindler objected. In 2000, they initiated court proceedings by which Terri's nutrition and hydration (which are not considered treatment under Florida law) were maintained. After some 14 cases of litigation, Terri's feeding tube was discontinued one last time and she died, of underlying natural causes, on March 31, 2005.

What is perhaps most interesting about all of this is that many viewed this as a right-to-die case, which it never really was; there was no preexisting living will or advance directive to enforce, although Michael contended that Terri had made oral statements. This was, in essence, a case regarding non-voluntary euthanasia of someone who had no capacity to consent. Secondly, as a matter of perspective, the case was in essence a family feud in which the husband (who had gone onto another relationship) sought to discontinue treatment and the parents sought to continue treatment. Both the Schindlers and Schiavo each alleged that the other was trying to get access to the money from Terri's trust. Further, both the Schindlers and Schiavo wrote books about Terri and about their experiences, published in 2005 and 2006, respectively. One might well argue that these activities were the civil equivalent of what the prosecutor in the final Kevorkian prosecution labeled a lack of dignity in the treatment of Tom Youk (by not closing Youk's mouth while the camera was still running). In any event, the case was not about voluntary euthanasia by a competent adult with a terminal illness; and while the initial PVS cases moved forward, the debate as to whether such individuals should be able to get access to assisted suicide or other means of hastened death was not what the Schiavo cases sought or did.

BAXTER V. STATE[16] (OF MONTANA) 2009: DECLARING A STATE CONSTITUTIONAL (NOT FEDERAL) RIGHT TO ASSISTED SUICIDE

Near the end of the post-millennial decade, in *Baxter v. State*, the Supreme Court of the state of Montana determined that terminally ill patients have a right, as a matter of state constitutional law, to seek—and receive—physician-assisted suicide. The civil litigation that led to the December 31, 2009, decision (over one year after Robert Baxter's death by natural causes due to his underlying lymphocytic leukemia, for which he had sought a variety of traditional treatments), decided as a matter of state constitutional law, as with Washington State's ballot initiative, took up the opportunity and invitation of the 1997 U.S. Supreme Court cases for states to decide how to deal with the questions relating to assisted suicide. The state high court's decision was made on the law, but was constructed in no small measure as a result of the American debate of the prior two decades. This construction may not have ensured the outcome of the seven-member panel's consideration, but did seek to avoid pitfalls of civil litigation and criminal liability that had been presented (to no ultimate avail, in the states where legal change was sought by methods other than ballot initiative). It did, however, circumvent a variety of politically perceived obstacles to a judicial ruling favoring assisted suicide, without subjecting it to further review by the U.S. Supreme Court.

Some of these political changes included a complete exclusion of federal constitutional provisions to due process and equal protection, which had failed at the U.S. Supreme Court level in 1997 in *Glucksberg* and *Quill*. By restricting the case to being one as a matter of state constitutional law, the Montana Supreme Court judges were unfettered in their determination, and secure in the knowledge that their decision would not be subject to further review.

In addition, the way in which the factual scenario was presented was heavily (and presently) patient-focused, whereas in *Quill*, the doctor had already been subject to a prosecution, albeit one that ended in his favor. The *Baxter* case also hearkened back to the first civil litigation—the 1994 Michigan *Hobbins* case, which many have been lost in no small measure due to being combined with criminal cases against Jack Kevorkian, a medical and legal pariah in the state.

The original named plaintiff was a retired truck driver, who had been treated for his leukemia with multiple rounds of chemotherapy, which became less effective as time progressed. A student of sociology might question whether the choice of a man—indeed, a man's man—as a named plaintiff patient was in no small measure to present a strong gender message to the court; this is in contrast to the 72 percent of Kevorkian patients who were women (although he was tried roughly equally for hastening the deaths of both men and women), and to the socially constructed perception that men succeed in suicide more than women, who attempt suicide more. This abstraction, while reaching back to Durkheim and forward to the 2000s, nonetheless may have contributed to the efficacy of a plaintiff patient.

Baxter's illness, unremitting and repeating, characterized by the court as "terminally ill" (which, by the time the Supreme Court contemplated its case, was without question, as Baxter had died from his illness, while Teresa Hobbins remained alive during and after the Michigan Supreme Court cases for some time). The case, brought in state court by Baxter, four physicians, and the interest group Compassion & Choices, challenged "the constitutionality of the application of Montana homicide statutes to physicians who provide aid in dying to mentally competent, terminally ill patients."[17] Thus, the state court was ruling solely on state law matters.

In addition, the complaint alleged—again solely as a matter of state law—"that patients have a right to die with death with dignity under the Montana Constitution . . . which addresses individual dignity and privacy."[18] The District Court (the lower court, or the court that first heard the case) agreed, and held that "a patient may use the assistance of his [*sic*] physician to obtain a prescription for a lethal dose of medication. The patient would then decide whether to self-administer the dose and cause his [*sic*] own death."[19] As an aside, this wording seems to underscore the importance of a man's decision (rather than a person's or an expression of "his/her," a subliminal gender cue that seems to support the gender politics abstraction previously discussed). Last, and most important to the question of how to treat physicians who assist in suicide (but not those who engage in medical euthanasia), "the District Court further held that the patient's right to die with dignity includes protection of the patient's physician from prosecution under the State's homicide statutes."[20]

In short order, the original decision allowed for the conduct of a Quill to be non-prosecutable, but not that of a Kevorkian. The methodology of the physician—providing the means, but not performing the final act—was held to be outcome determinative as to whether a doctor could be criminally liable. (Indeed, under this scenario, Pou would not have been excused.) Additionally, the lower court's decision, presented for review by the Supreme Court, tracked the sort of conduct that had been made legal in Oregon by Measure 16, and upheld by the U.S. Supreme Court, and the conduct most recently voted within lawful doctor-patient relations in Washington State.

The Montana Supreme Court opened with its own political statement, "the proposition that suicide is not a crime under *Montana* law."[21] The law pertaining to suicide and its decriminalization was heavily discussed by the legal theorists of the 1950s, and decriminalized thereafter in both England and Wales, as well as the American states in which suicide was a crime. However, as a statement of public policy, and using the English *Suicide Act 1961* as a theoretical foil, assisting in a suicide was largely viewed as criminal. As in the 1984 Michigan case of *People v. Campbell*, which was one of the cases contemplated during the 1990s Kevorkian prosecutions, promoting a suicide remained a crime after suicide was decriminalized; and of course, the Michigan legislature enacted repeated bans on assisting in a suicide as a result of Kevorkian's activities during the 1990s.

However, just as Washington State's evolution from the initial failed initiative in 1991 through the failed federal suit that resulted in the unsuccessful 1997 Supreme Court challenge, served to inform its experience in setting a voter initiative that would pass muster with its populace (by modeling Oregon's, which successfully upheld the U.S. Supreme Court challenge), the Montana Supreme Court had the benefit of considering cases in civil litigation and criminal prosecutions. Hence, the majority took the approach of further delineation, immediately removing what it politically couched, throughout the decision, as "aid in dying," from suicide as well as homicide. In a related vein, Justice Nelson, in a special concurrence (agreement with the result, but distinguishing commentary), took the question of taxonomy and definition further, which underscores the ongoing evolution of vocabulary and issues in the euthanasia and assisted-suicide debate. In his discussion about when a terminally ill person may "expect death within a relatively short period of time,"[22] the jurist noted that "in choosing this language, I purposely eschew bright-line tests or rigid timeframes . . . [since] what is relatively short varies from person to person."[23] This returns to the question of there being more than one definition of terms, as well as time frame, as Judge Nelson goes on to discuss when considering what constitutes suffering.

The Montana decision was, however, largely focused upon whether the patient "consented," and defined consent in terms of capacity, rather than embracing the traditional view that one cannot consent to their own death as a matter of Anglo-American law, or that such might be against the interest of the state (indeed, the court went out of its way to say that "we similarly find no indication in Montana statutes that physician aid in dying is against public policy").[24] Rather than issue a blanket ban on consenting to assisted suicide (as did Michigan), the Montana court, relying upon state statute, "if the State prosecutes a physician for providing aid in dying to a mentally competent, terminally ill adult patient who consented to such aid, the physician may be shielded from liability pursuant to the consent statute [although that is] only effective if none of the statutory exceptions to consent applies."[25]

Ultimately, the Montana Supreme Court determined that "a physician who aids a terminally ill patient in dying is not directly involved in the final decision or the final act. He or she only provides a means by which a terminally ill patient can give *himself* [*sic*, emphasis in original] can give effect to his [*sic*] life-ending decision, or not, as the case may be."[26] In keeping with the gendered language raised earlier, it is interesting to note that a doctor was viewed as being of either gender, while the patient male; although this patient was male, one might have assumed that future patients would be both male and female, in varying proportions or equally proportionate. However, the effect of the decision should not be lost—the conduct of a Quill was ratified as consistent with Montana state law, whereas illegal in New York (still) and Michigan (as legislated, regarding criminal liability for providing the means to a patient).

Perhaps the single most important depiction of the view of the majority was expressed when it said that:

"Each stage of the physician-patient interaction is private, civil, and compassionate. The physician and terminally ill patient work together to create a means by which the patient can be in control of his [*sic*] own mortality. The patient's subsequent private decision whether to take the medicine does not breach public peace or endanger others."[27]

This, in combination with the fact that "further, the legislature criminalized the failure to follow a patient's end-of-life instructions."[28]

In holding that patients have the right to "aid in dying" (or assisted suicide) from physicians, who would be saved harmless from criminal liability, the majority did leave a gap, as noted in the concurrence of Justice Warner. Specifically, while commending the majority for creating a holding based upon statutory law (rather than broader principles), the jurist as critical that "[t]he logic of the Court's opinion is not necessarily limited to physicians. In [Judge Warner's] view, the citizens of Montana have the right to have their legislature step up to the plate and squarely face the question presented in this case."[29]

This suggests that perhaps Judge Warner might have taken the logic further and permitted those who assist in ways other than by prescribing the lethal prescription. This criticism may be couched in terms of Oregon, and the 2010 HBO documentary *How to Die In Oregon*, in which doctors wrote prescriptions that volunteers from Compassion & Choices (an amicus in the Montana case) and family members helped prepare, and helped patients to self-administer. The Montana holding would exclude these individuals and leave them within the ambit of criminal liability, notwithstanding a lawfully written physician's prescription.

In dissent, Judge Hegel also expressed concern as to the family, partners and friends of the patient. In specific, jurist wrote that the:

Court's approach is . . . disconcerting when considering the ambiguity this [majority] opinion [and thereafter state law] will bring for those who are not physicians. Physician assistants, nurse practitioners, nurses, friends, and family do not qualify as physicians. But the will all undoubtedly be involved to varying degrees in the process of physician-assisted suicide. Yet the Court's public policy reasoning is based upon the role of the physician. The net result of the decision, whether intended or not, is to leave "non-physicians" with the question of whether the decision premised upon a physician-based policy will apply to them as well.[30]

Judge Hegel in fact used this logic as additional support for Judge Rice's dissent (in which the jurist argued that the lower court should be reversed in its entirety, for violating the prohibitions on suicide and the previously standing statute of 114 years). However, Judge Hegel in fact points to a differentiation between the legislated Oregon and Washington Death with Dignity Acts (which allow for such presence and family and friends) and what Montana would allow for as a result of the *Baxter* decision. In effect, Judge Hegel issued an additional criticism to the Rice dissent, to the effect that patients will be further isolated by the decision to allow for a prescription from a physician, but not loving support of family and friends at the time that the patient chooses to use the prescription. While there have yet to be studies in the United State similar to the Dutch study of Swarte et al., the HBO Oregon documentary would suggest that families fare better when they are able to say goodbye, and also that patients seem to feel more comfortable (although one must note that the Oregon film was particularly sympathetic to the issue).

Thus, one might conclude that the Montana decision, while well considered in terms of the legal challenge, was unable to provide for the various safety valves that the Oregon and Washington assisted-suicide laws provide for with procedural safeguards and regulatory mechanisms that include additional doctors, psychological consultations, and family and social support.

WASHINGTON STATE: A DEATH WITH DIGNITY ACT FOR THE 2000s ALLOWING FOR DEATH TOURISM

The medical euthanasia and physician-assisted suicide debate was most obviously going to carry on in Washington State in the 2000s, once the matter was settled in the 1990s by legislature (as against assisted suicide) in Michigan and by ballot initiative (in favor of assisted suicide) in Oregon, the latter with procedural affirmation by the U.S. Supreme Court in 2005. It was indeed a logical progression for there to be a further ballot initiative in Washington following the U.S. Supreme Court's decision to uphold the Oregon law, on narrow procedural grounds, rather than considering more sweeping issues.

In 2008, Washington State saw Initiative 1000 on its ballot, which passed on November 4, 2008 (the same day that Jack Kevorkian lost a bid for elective office in Michigan). The act passed by a margin of 57.8 percent to 42.2 percent, and was codified as RCW 70.245 *et seq*., with an effective date of March 5, 2009. In most ways, the Washington Death with Dignity Act was patterned upon the Oregon Death with Dignity Act, and also regarded terminal illnesses with a projected life expectancy of six months or less.

As with the Oregon law, Washington State's law provided for public reporting. Between March 5, 2009 (when the act became law), and December 31, 2009, 53 pharmacists issued prescriptions to 63 individuals; 37 individuals died, 36 of whom died after ingesting medication (and seven of whom died without using the medication).[31] The age ranged from 48 to 95; 79 percent had cancer, 9 percent had neuro-degenerative illness (including ALS), and 12 percent had respiratory disease or other illnesses.[32] The social attributes were that 98 percent were white, non-Hispanic; 55 percent were male and 45 percent were female; and 61 percent had college educations.[33] Some 94 percent were at home, and 72 percent were enrolled in hospice care at the time they took the medication and died.[34] In 2010, 68 physicians issued prescriptions to 87 individuals, 51 of whom died after taking the medication and 35 of whom died without having taken the medication.[35] Other statistics were similar to 2009, with an age range of 52–99 years, 78 percent cancer victims, 10 percent patients with neuro-degenerative illness, and 12 percent with heart disease or other illness.[36] Some 95 percent were white, non-Hispanic, 51 percent married. and 62 percent college educated.[37] A full 90 percent of those who availed themselves of the pharmaceuticals were at home, and 84 percent were enrolled in hospice when they took the lethal medication.[38] An even 50 percent were male and likewise female.[39] End-of-life concerns included loss of autonomy (90 percent), loss of ability to engage in activities that made life enjoyable (87 percent), loss of dignity (64 percent), loss of control of bodily functions (64 percent) concerns about being a burden on family/friends/caregivers (28 percent), and, far down the list, concerns about inadequate pain control (36 percent) and those of the financial implications of treatment (31 percent).[40]

Similarities aside, one example of a distinction between the Oregon and Washington laws regarded residency. The Washington State law provided that:

> Only requests made by Washington state residents under this chapter may be granted. Factors demonstrating Washington state residency include but are not limited to:
> (1) Possession of a Washington state driver's license;
> (2) Registration to vote in Washington state; or
> (3) Evidence that the person owns or leases property in Washington state.[41]

This provision, while requiring Washington State residency, did not require a specific time period in which residency must be required, as Oregon did, and as Nancy Niedzielski, a Seattle woman who moved the (successful) Washington I-1000 campaign forward, experienced. Indeed, Niedzielski became an activist for the Washington initiative specifically after her husband was deemed not eligible under the Oregon law, because he could not establish legal residency between the time he was diagnosed with brain and spinal cord cancer and the time of his death the next year, as documented with her interviews in the HBO film, *How to Die in Oregon*.

While the removal of a time requirement relating to residency was intended to make for an allowable death with dignity for those who may move from another locale, a fair criticism of this provision is that it will allow for what is in effect death tourism. As a two-year old law at the time of this writing, it is perhaps too early to determine whether death tourism will increase as a result of the Washington State law (as in contrast to the Oregon law, which does have a mandatory minimum time attached to the residency requirement). As something of an irony, this more progressive residency requirement in the Washington State law is not an affront to St. John-Stevas and the questions of the sanctity of life, but rather might be pointed to by those who embrace Yale Kamisar's construct that the slim edge of the wedge can lead to a slippery slope. In particular, whether Washingtonians refined the Oregon law for patients or whether the state has taken a step forward into death tourism is an argument which may be based upon the vocabulary and politics of the debater, showing yet another way in which the politics of assisted suicide are subject to historical perspective and analysis.

However, it is noteworthy that during roughly the same time period that Washington State has had assisted suicide (with potential for death tourism), a number of individuals have specifically sought death tourism in Switzerland's Dignitas clinic. Debbie Purdy, a British woman from Bradford who had a diagnosis of multiple sclerosis, challenged the laws of England and Wales, with an argument that melded Baxter's claims against Montana with those of death tourism. Remaining alive at the time of this writing, during 2009, she successfully argued to the Law Lords that she had a right to know whether her husband would be prosecuted for assisting in a suicide if he furthered her travel to the Swiss Dignitas clinic to take an overdose of barbiturates, in a direct challenge to the *Suicide Act 1961*. Purdy's contention was that her husband Omar would be at risk of prosecution under section 2, which provides for penalties up to 14 years of people who "aid, abet, counsel or procure the suicide of another."[42] Thus, while section 1 provided for the abolition of suicide as a crime, helping to provide the means or opportunity remained so (although nobody had been taken to trial in the United Kingdom. in that regard). Of no small matter was that Purdy was seeking to commit suicide in Switzerland (not the United Kingdom), and thus, at least theoretically, outside the jurisdictional ambit of the law.

In Purdy's challenge, she contended that if her husband had no guarantee of not being prosecuted, she would have to make the trip sooner, rather than later, and while she could still travel without assistance. In this regard, there is again a hearkening back to the Dutch study of Swarte et al., in that family being able to say goodbye was regarded as an issue and, implicitly, that the supportive company of family and friends was essential, as in Oregon; that said, the case highlighted again that Oregon had a residency requirement, which some would-be assisted-suicide patients were finding cumbersome, problematic, and obstructive. As a result of the Purdy case and similar situations, Keir Starmer, the director of public prosecutions, issued a list of conditions under which it would be unlikely to prosecute those who helped friends or family members kill themselves. The director of public prosecutions "listed 13 factors that could influence the authorities not to prosecute [which included] the person aiding a suicide being motivated by compassion; the deceased clearly wanting to die; and the deceased being terminally ill, being severely physically disabled or suffering from an incurable degenerative disease."[43] In the same document, 16 factors that would promote prosecutions were listed, including "the deceased being under 18, mentally handicapped or not sure about his or her wishes, or not being seriously disabled, being terminally ill or suffering from a degenerative disease. They also include the person aiding the suicide pressing someone into it or being motivated by personal gain."[44] The DPP's guidelines were a direct response to the comment of the Law Lords in *R. (on the application of Purdy) (Appellant) v. Director of Public Prosecutions (Respondent)*,[45] the previous July that while it was up to Parliament to enact a law, the director of public prosecutions was required to clarify factors and standards relating to law enforcement.

The British approach ironically supports death tourism (as does the Washington State law), and also goes further, by supporting assisted suicide in disease that is not necessarily terminal, i.e., degenerative. This indicates that the wedge may be perhaps widening, albeit with compassionate motivation.

CONCLUSION

During the post-millennial decade, a variety of criminal and civil cases and new legislation (as well as the 2006 Supreme Court decision upholding the Oregon law, more considered in this project as an extension of the 1990s ballot initiative approving death with dignity by assisted suicide). As something of an irony, these included at least two states (Georgia and Louisiana) in which the law was deliberately misapplied, by the plea bargain to the fictive assisted suicide in Carr's mercy killing in Georgia, and by the grand jury's refusal to indict Dr. Pou after her role in the euthanasia of patients in post-Katrina New Orleans. These cases had more to do with assisted suicide and euthanasia than did the Schiavo case in Florida, which in current socio-legal and medical parlance was more about a family dispute between parents and a husband in ending care of a patient in a persistent vegetative state.

Washington State, on the other hand, provided full closure from its early 1990s debate (and failed ballot initiative) regarding medical euthanasia, through to the civil litigation that was ultimately unsuccessful in the U.S. Supreme Court in the late part of the decade, and into its own robust assisted-suicide legislation, following the example (and survival) of the Oregon law. As with Oregon, the early reports suggest that not all who qualify for assisted suicide avail themselves (in contrast to the Michigan Kevorkian cases) and that the gender and health statistics suggest a fairly used law that has the benefit of a robust regulatory system of safeguards. While Montana has preliminarily gone the route of judicial declaration, the decision itself reflected that more discussion and debate will have to be had to ensure both patient sensitivity and provider or family protections.

It was a busy decade, for which the previous century's legal theorists, civil and criminal cases, and legislative efforts all came into play, as well as the questions of criminal liability for doctors and others who engage in medical euthanasia, mercy killing, and assisted suicide.

NOTES

1. Stirgus, Eric, "Woman Who Killed Sons Leaves Prison: Family, Friends Celebrate Release," *Atlanta Journal-Constitution*, March 2, 2004, D3.

2. Fletcher, George P., *Rethinking Criminal Law* (Oxford: Oxford University Press, 2000), 377.

3. *Id.*

4. Code 1981, § 16-5-5, enacted by Ga. L. 1994, p. 1370, § 1; Ga. L. 2007, p. 133, § 5/HB 24.

5. Scott, J., "Bond Delayed for Accused Killer of Sons," Atlanta Journal-Constitution, June 21, 2002, C1. Retrieved from http://search.proquest.com/docview/336902175?accountid=14055.

6. Fink, Sheri, "The Deadly Choices at Memorial," *New York Times Sunday Magazine*, August 30, 2009, 28.

7. *Id.*

8. Fletcher, *op. cit.*, 823.

9. Pakes, Francis, *Comparative Criminal Justice* (Devon, UK: Willan Publishing, 2004), 58.

10. Fink, *op. cit.*

11. *Id.*

12. Wolfson, Jay, "Foreword," in Arthur L. Caplan, James J. McCartney, and Dominic A. Sisti, eds., *The Case of Terri Schiavo: Ethics at the End of Life* (Amherst, NY: Prometheus Books, 2006), 13.

13. *Id.*, 14.

14. *Id.*

15. *Id.*, 15.

16. *Baxter v. State*, 354 Mont. 234 (Mont. 2009).

17. *Id.*, 354 Mont. at 238.
18. *Id.*
19. *Id.*
20. *Id.*
21. *Id.*, 239.
22. *Id.*, 256.
23. *Id.*
24. *Id.*, 243.
25. *Id.*, 239–40.
26. *Id.*, 242.
27. *Id.*
28. *Id.*, 246.
29. *Id.*, 252.
30. *Id.*, 278.
31. Washington State Department of Health 2009 Death with Dignity Act Report.
32. *Id.*
33. *Id.*
34. *Id.*
35. Washington State Department of Health 2010 Death with Dignity Act Report.
36. *Id.*
37. *Id.*
38. *Id.*
39. *Id.*
40. *Id.*
41. *RCW* 70.245.130.
42. *The Suicide Act 1961*, Section 2.
43. Lyhall, Sarah, "Guidelines in England for Assisted Suicide," *New York Times*, September 23, 2009.
44. *Id.*
45. [2009] UKHL 45.

Concluding Commentary

INTRODUCTION

This chapter is deliberately not entitled "Conclusion," because the debates surrounding acts of medical euthanasia and assisted suicide continue to carry on. Attempting to keep writings current with developments has, over the past 20 years, proven to be a daunting task. On more than one occasion, articles and writings that were intended to be contemporary were deemed historical in nature even before they were printed. For example, in 1992, the author's master's dissertation, due in September, had to be delayed until October, to accommodate for (among other things) inclusion of the trial of Winchester doctor Nigel Cox, who was ultimately convicted of attempted murder (due to lack of a body upon which to perform a final autopsy). Barely two years later, in 1994, a piece entitled "Euthanasia and Assisted Suicide: Are Doctors Duties When Following Patients' Orders a Bitter Pill to Swallow?" had to be repeatedly removed as late as the galley stage, to add information about Jack Kevorkian's activities in Michigan. The field work for the doctoral dissertation that preceded the instant book treatment was repeatedly reopened, until Kevorkian's final conviction in 1999. The author attempted to retrieve the original doctoral dissertation filed, when within a few short weeks, Washington State voters approved a ballot initiative allowing for physician-assisted suicide, ironically on the same day that Jack Kevorkian lost a bid for public office.

So it is with little doubt that this author predicts that, by the time this book is in press, it will no longer be an up-to-the-minute consideration of the ongoing debate. In fact, during the final drafting of this book project, the legislature in Vermont began to consider whether or not assisted suicide should be legal. It was perhaps for this reason that the Sundance Award–winning best documentary of 2010, *How to Die in Oregon*, had its public premiere in Vermont (rather than the more traditionally expected home state of Oregon). While not yet a historical perspective, although it may be so by the time that the book is printed, it is noteworthy that the Vermont legislation is entitled "An Act Relating to Patient Choice and Control at the End of Life." Because Vermont has a two-year legislative session, the possibility of physician-assisted (prescribed) suicide remains on the legislative agenda in 2012.

Also pending in 2012 is a Massachusetts initiative for a Death with Dignity Act, which may appear on the November 2012 ballot. The initiative will appear if the legislature does not create legislation sooner, which has been hinted at.

Georgia, where Carol Carr was given a plea deal, allowing her to plead to assisting in a suicide rather than face manslaughter charges for shooting her two sons to death as they lay, afflicted with Huntington's disease, in a nursing home, has also been a geographic area of activity in 2012. In fact, the Georgia Supreme Court determined that a law prohibiting assisted suicide was unconstitutional. As a result of this, four people, including Dr. Lawrence Egbert, who were charged with violating the assisted-suicide law (among other things) have had pending charges against them dismissed.

Hence, as an introduction to this series of concluding remarks is the prediction that the commentary will be outdated by the date of printing (if not before). The ebb and flow of the tide of legislative enactments, civil litigation, and criminal prosecution regarding assisted suicide (and the more secretive practice, other than that of Kevorkian, of medical euthanasia) will have proceeded (possibly in both directions) and receded. For some, that may make this book a truly historical perspective, refracted through the particular lens of criminal law and criminal justice policy; for others, that may render this writing an outdated one, if another way of examining concluding commentary is not utilized.

Thus, approaching the concluding commentary by way of consideration of thematic issues, interwoven with temporal timelines, invites readers to draw their own conclusions. There may be those who might say this is a cheap way out of giving an actual conclusion pro or con (an actual accusation at the dissertation defense, where every substantive question was handily answered, but actual opinion reserved in the absolute). In teaching on this topic, there are students who pry for answers and find their ordinarily forthcoming professor surprisingly reluctant to give an opinion. Perhaps a great compliment was when a seasoned criminal lawyer and judicial officer asked if this author had an opinion, and I said, "Yes," in a flat tone; then said

"No" to the follow-up question of whether I would tell him (at a table filled with lawyers well skilled at interrogation techniques) what my opinion was. The implicit compliment was in that none of the people at the event were quite certain of what this author's personal opinion was. That approaches the Baroness Flather's experience of having people come up to her in the corridors of the House of Lords and variously comment that of course she was in favor of medical euthanasia, or, in the alternative, that she was obviously in opposition to medical euthanasia.

The purpose of this book was never to persuade a reader to be in favor or to be in opposition of either medical euthanasia or physician-assisted suicide. Rather, the purpose was to promote thematic thinking along a timeline. While perhaps kaleidoscopic, that is how the concluding commentary hereinafter should be considered.

A GENERAL COMMENT ON THE MOVE TO LEGALIZE MEDICALLY HASTENED DEATH

One immediate comment, which precedes all others temporally, is that the initial movement to legalize medical euthanasia in the early 1900s preceded most efforts to decriminalize suicide (generally) and to legalize medical assistance in the suicide of competent, terminally ill adults with decision-making capacity. This, it seems, is largely overlooked when there are discussions of patient autonomy and self-determination, whether one is in favor of or opposed to patients choosing to die with assistance (or, alternatively, having doctors participate in hastened death of patients).

That the initial movement in the early 1900s was designed to have doctors be the final arbiters and the final actors in hastened death of patients who were terminally ill, or in chronic pain, or suffering from a debilitating and uncorrectable condition, does not mitigate against the late 1990s conduct of Jack "Dr. Death" Kevorkian, but does perhaps offer a longitudinal view of medical culture. This culture historically has been one where (in modern medicine) a doctor has been in "control," to use Kevorkian's word, for better or for worse. However, as will be amplified, during the 1990s a second model, with the patient as consumer of medical services, and one with control of the time and manner of death, if not the fact of dying, arose.

SPECIFIC COMMENTARY ON THE CHANGING ROLE OF DOCTORS IN THE MEDICAL EUTHANASIA DEBATE AS A CHRONOLOGICAL AND TOPICAL MATTER

Between the model of doctors as all-knowing, all-powerful elites and the rise of the patient as consumer/doctor as provider model (one at which many doctors take umbrage), the criminal law (as well as medical culture) and criminal justice policy began to contemplate what the role of doctors at the end of life should be. More to the point, what began developing was a socio-legal and medico-legal culture of what kind of conduct would be considered to be so deviant as to be beyond the scope of a reasonable doctor-patient relationship and appropriate (or, more to the point, inappropriate) physician conduct. Placed in context, that which was seemingly typified (giving overdoses of morphine to cancer patients) began to be considered under extraordinary circumstances.

First, Lord Dawson of Penn's medical euthanasia of King George V, with the consent of his wife the Queen, showed a world a number of extraordinary events. First, the death of the King by these illegal means was in and of itself stunning. Second, the fact that the King's hastened death was actually timed so as to be newsworthy on a particular date and time reflects an intentionality and planning that Lord Dawson was not held to account for (but later in the century, Jack Kevorkian would be, in no small part for that very reason). Third, that the information about the King's medical euthanasia was withheld for some 50 years amply demonstrates the way in which doctors felt that they had to practice in the shadows; regardless of whether one views this death as a private family matter of a terminally ill man, or whether one views the death as a one of public impact by change of monarch, the conspiracy of silence was consistent with what others earlier in the century, with less famous patients, described. Fourth, the fact that Lord Dawson's own biographer continued to withhold this information underscores this. Fifth, and perhaps most compelling, is the fact that Lord Dawson was a proponent of medical euthanasia, but an opponent of the regulatory mechanisms under consideration in the 1936 House of Lords debates, later that year (even leaving aside the conflict of interest matter as an unnamed defendant or coconspirator, this again underscores the seeming sense of entitlement that doctors had and their lack of perception that they were within the reasonable ambit of the criminal law for violating it).

Whereas discussion of changing the law regarding medical euthanasia started in the late 1800s and early 1900s, the first actual legal changes in Anglo-American law emerged in the 1950s (after yet another failed effort to get a medical euthanasia law passed in England in 1950, and the failed effort in Nebraska in the late 1930s). The irony of the first case to be acknowledged by legal theorists as having an impact upon the medical euthanasia debate, that of Dr. Hermann Sander, is that the New Hampshire defense team did not argue in favor of mercy killing or advocate a theory of medical euthanasia. Rather, the defense theory of the case, that the patient was already dead when Sander injected her with air bubbles (which would unquestionably have killed her, assuming a live person) and witnesses who testified to the effect of a man on the brink of emotional collapse due to deaths in personal and professional circles, all pointed to a theory of legal impossibility (with a sufficient hint of extreme emotional disturbance to mitigate against a conviction under the charges and the theory of the prosecution's case). Importantly, the patient's family and community supported him, offering him a parade upon his acquittal and a return to full practice. This would prove to be predictive of the English case and response to the Nigel Cox prosecution, where he was tried in a winking prosecution for attempted murder (rather than murder), supported by the "victim's" family, given minimal negative sanctions and remedial palliative care training by the Greater Medical Council as a penalty, and sent back to practice as a rheumatologist.

Intriguingly, the legal theorists of the late 1950s cited the Sander case, but yet each one failed (or took a deliberate decision not) to discuss the British case against Dr. John Bodkin Adams. The reason that this is an interesting note to comment upon is that this case—not the Sander case—was the one that provided a game-changing medical and legal defense of "double effect." This exemption or out clause meant that doctors who prescribed morphine (or other drugs) in large quantities, with the secret intention of ending patient suffering (also known as euthanizing or killing, by virtue of the word "ending"), would now have a clear defense, whether factual (the primary intent was to mitigate pain, though it secondarily shortened the patient's life) or fictive (the actual intent was to end the patient's life in a pain-free way). While Adams was a disgraced doctor, stripped of licensure for similar bad acts, the jury instruction issued by Lord Devlin became a standard charge in medicolegal proceedings. The phrase "double effect" found its way into medical-euthanasia and assisted-suicide parlance before the phrase "to Kevorkian" did so at the end of the century, though most people who use the former never heard of either Adams or Devlin, the originator of the exception.

While the legal theorists of the 1950s debated whether or not, and why or why not, medical euthanasia should be allowed, the actual (unsuccessful) criminal prosecutions planted a seed for the changing role of doctors. Whereas in the early 1900s, the question might have been presented that medical technology allowed people to live longer, and thus more likely to die of painful degenerative or chronic diseases, the question during that arose during the 1950s appeared to be how the changes in the roles and conduct of doctors should be treated by the social institutions involved in law and medicine, and by societal segments such as families.

This all said, as medical and legal priorities and protocols developed, there were matters that were clearly considered extrinsic to the appropriate role of doctors. One prime example of this was Jack Kevorkian, who may have ultimately been convicted for his public appearance on the CBS newsmagazine

60 Minutes, on which he narrated a tape of his euthanasia of Tom Youk. Showing the moment of the patient's death, announcing that euthanasia was "better control" for the (then out of license) doctor, and engaging in matters that seemed to subvert the patient's dignity (such as not closing Tom Youk's mouth after he was dead) appeared to be offensive to the jury. Each one of these individual parts of the episode could have been construed as beyond the role of a doctor, even under the most permissive of circumstances allowing for medical intervention to hasten death. Also offensive to the jury (though not broadcast to a wider public) was likely to have been the image of Jack Kevorkian straddling Tom Youk's wheelchair to adjust the intravenous that would deliver a lethal cocktail of sleep medication, muscle paralytic, and heart-stopping drugs akin to the drug delivery system of lethal injections in death penalty cases (which were, themselves, subject to litigation in the U.S. Supreme Court as to whether they were a medically cruel and inhuman punishment, during the decade following Kevorkian's conviction).

A second example of this, ironically discovered at the same time that Kevorkian was being tried and convicted, was Britain's Dr. Harold Shipman. This formerly respected doctor was convicted in 2000 of murdering some 15 of his patients by opiate overdoses. Unlike Kevorkian (who took no money), but reminiscent of Bodkin Adams (who was not convicted), Shipman was named as a beneficiary by some decedents (the true number of which was ultimately believed to have numbered some 218, which make Britons deem him the most prolific serial killer in their history, in contrast to Kevorkian's 130+ assisted suicides and acts of euthanasia, some of which went uncharged).

What Shipman and Kevorkian had in common was a severe disproportionate number of female to male patients/clients/victims, with Kevorkian's at 72 percent and Shipman's hovering around 80 percent. This also implicates the role of doctors, for even if there is an appropriate role of doctors to provide an easy exit from pain and suffering, then such gendered disparities must be immediately suspect, and in stark contrast to minimal gender gaps in Oregon and Washington State, where the role of doctors is to simply provide a prescription for a lethal dose for a patient, and only after very strict protocols have been fully complied with. Perhaps this is why those laws passed, whereas more aggressive ballot initiatives in the early 1990s failed in California and Washington State.

THE MOVE TO DECRIMINALIZE SUICIDE AND THE GROUNDWORK FOR ASSISTED SUICIDE

If the 1950s were a watershed decade for changes in the role of doctors in the death trajectory, the 1960s was likewise a decade for changes in how suicide would be treated. While suicide was decriminalized state by state (and not all during the 1960s) in the United States, the *Suicide Act 1961* demonstrated how the English, on a national level, were changing legal and societal approaches to suicide. No longer was suicide to be a crime, with forfeiture visited upon the surviving family. That this penalization was removed from families, along with the decriminalization of the suicidant him- or herself, prepared the groundwork for what would follow in doctor prosecutions of the 1990s, and the movements to legally legitimate assisted suicide of terminally ill patients in Anglo-American culture.

The legal and cultural destigmitzation of suicide, following on the temporal tail of the so-called right-to-die cases of Quinlan and Cruzan, perhaps inexorably led to a cohort of people, of patients, who did not want to have excessive death trajectories prolonged by extraordinary measures becoming available in medical technology. The Quill case, of a hospice physician approached by a patient seeking a prescription for a lethal dose to self-administer, showed that the 1990s would be marked by active patients (and, later in the decade, activistic patients), who would sue their various states, unsuccessfully, for the right to what Quill provided quietly and privately, but then wrote about in the premier medical journal of the country. It is likely that many patients made similar requests of doctors across the country, and that many doctors complied with patient requests in private.

The turning point was when Quill took the decision to see around the bend, to create the bend itself in the medical community, at knowing risk to himself in the legal and social communities. That Quill was not indicted suggested that the grand jury was taking a step past the step taken by the trial jury in the Sander case of the 1950s—that rather than seek a fictive defense, the social world reflected by the grand jury was prepared to put a terminally ill patient's choices as ranking higher than a doctor's actions in terms of allocating responsibility from the beginning. This is demonstrated by the failure to indict (rather than an acquittal at trial), and created in Quill an activist who (with others) took federal suit seeking a declaration that assisting in the suicide of a terminally ill patient was to be considered constitutionally proper and within the scope of medical care, if no cure could be had and if pain and suffering of the patient had become intractable.

Unintentionally, and ironically, Quill's prescription script and Jack Kevorkian's first cases were within a year of each other. Speaking with either of the two makes (or made, as Kevorkian is recently deceased) it clear that not only had they not been in cooperative (or collusive) interaction with each other, but rather that they had individual paths to follow. For Quill, it was patient empowerment, with the objective of giving the patient a safety net by which it was possible to stay alive for a longer period of time. For Kevorkian, it was doctor empowerment, with an ultimate objective of giving the doctor more (and ultimate) control over the process, never intended to prolong a patient's life. These unquestionably laid the groundwork for a decade replete with assisted-suicide debates in the medical community (that continues to embrace Quill, though it ejected Kevorkian) and the legal community (through the Quill/Vacco suits in the federal system and the Kevorkian/ Hobbins cases in Michigan) during the 1990s.

MOVING TOWARD (OR AWAY FROM) PHYSICIAN-ASSISTED SUICIDE

The early 1990s activities of Quill and Kevorkian led, on parallel paths, to different results. While some have immediately credited (or discredited) Kevorkian, and the schedule of his activities, with resulting in the defeat of the Washington State and California ballot initiatives in the early 1990s, most have credited him with bringing attention to the social forefront. This was, ironically, by the same media campaign conduct that ultimately led to Michigan criminalizing assisted suicide and imposing specific bans upon the conduct, and which also led ultimately to Kevorkian's self-precipitated trial to conviction for the Youk euthanasia (not assisted suicide, of which Kevorkian was always acquitted).

Hospice physician Quill's activities, and scholarly writing and excellent presentation as a named plaintiff, led the law forward (despite his defeat in the U.S. Supreme Court) by bringing out the possibility of an alternative to medical euthanasia, one that could be controlled by patients. Indeed, this is what the legislation ultimately passed in Oregon and Washington State provided for, and under very strict medico-legal protocols and safeguards.

In time, other states may go on to further legislate for lawful, or against unlawful, assisted suicide. Indeed, as mentioned at the beginning of this conclusory commentary, such an effort is under way in Vermont at the time of the final edit of this book.

POSTER CHILDREN: JACK "DR. DEATH" KEVORKIAN, TERRI SCHIAVO, AND THE MEDIA

The cover of this book has images of two of the most famous faces in the euthanasia and assisted-suicide debates—Jack "Dr. Death" Kevorkian and Terri Schiavo. Both have purportedly represented the best and the worst of the debate, the former by what was almost certainly a methodical plan (which included systematically contacting and communicating with the press), the latter unconscious of the family feud she precipitated once in a persistent vegetative state, let alone the national press and controversy.

Some viewed Kevorkian as a patient advocate crusader, others as the most prolific and successful serial killer of the modern time—one who found a way to legitimate ghoulish and macabre deviance by way of medical culture, all while managing to challenge the Michigan legal system repeatedly and variously. There is no question that Kevorkian captured the public imagination of those in favor of and those opposed to assisted suicide (and also medical euthanasia).

In the Kevorkian cases, testimony of the families, mandated in the early cases (in efforts to prove the element of intent) and banned in the final case to criminal conviction (in a successful effort to preclude evidence of Tom Youk's pain and suffering), brought the anguish of the families of the decedents into the light of day in court, and the hues of television in nightly news (as well as press). These cases provided a shift from cases of years gone by, in which family members were often the final actors in mercy-killing cases, to the role of families as a support structure for those seeking assisted death from Kevorkian. As time went by, families went from being "mere" witnesses to becoming assisted-death advocates, as did Carol Peonesch (who was the daughter of Merian Frederick, of whose assisted suicide Kevorkian was acquitted, and who was one of the prime movers of the failed effort of Michigan's ballot initiative to legalize assisted suicide) and the family of Tom Youk, who participated (in favor of both Kevorkian and assisted suicide) in the 1998 *60 Minutes* "Death by Doctor" segment that ultimately led to his conviction. These families ultimately began to bring their personal anguish about the ordinarily private matter of death of a loved one into the public sphere.

Over the course of a decade, Kevorkian's image and name both became inextricably intertwined with assisted death (more usually assisted suicide, but also medical euthanasia). Given Kevorkian's proclivities for publicity, most physicians distanced themselves from a man considered to be a publicity seeker, rather than one seeking to deal with the private matter of death and dying.

Over the course of the following decade, Terri Schiavo's image became the face of many who were both in favor and in opposition of the right to die by withdrawal of medical treatment. However, films of Schiavo were perhaps less about the right to refuse treatment or the best interests of the patient, and more about family who were press savvy, appearing on talk shows, hiring press agents, and writing books about their family feud. The fact that Schiavo's parents and husband each accused the other of having a financial motive appears to have been less about how to treat (or withdraw treatment from) patients in a persistent vegetative state and more about a private family fight deliberately brought into the public eye. Like Kevorkian, Terri Schiavo captured the imagination of the medical, legal, and social worlds of the general public, by way of massive media attention. However, this family was not so much reminiscent of the anguished Quinlan (whose photograph in life was the iconic image, rather than images of her in a persistent vegetative state, like Schiavo) and Cruzan families, seeking simply to resolve matters in the best way for their family member; but rather of more complicated issues, and thus perhaps not the most suited poster child for the already complex personal trouble of how to deal with a family member or beloved in a persistent vegetative state.

At the end of the day, a poster child may evoke a folk hero, a media darling (voluntarily or not), a folk devil, or a folk tragedy. Rarely, if ever, is there a uniform viewing of the poster child in political or social terms, a danger seen in these cases.

ON REIGNING SUPREME: U.S. AND STATE SUPREME COURT CASES

In 1997, in ruling against assisted suicide in the paired Washington and New York cases, the U.S. Supreme Court pointedly left the issue of how to regulate assisted suicide to the states. While some of the justices seemed in favor, and others clearly opposed, to assisted suicide, the paired decisions ultimately determined that there was no federal due process constitutional right or equal protection constitutional claim for a patient to die with assistance, and hence no protection from civil or criminal liability for a physician who provided assistance in death. That said, the Supreme Court made the point that states could decide how to effect such a decision.

It is likely that the justices, individually and collectively, knew that they would again see the issue rise before them. They may already have been contemplating the likelihood that they were going to review the lawfulness of the 1994 Oregon Measure 16, and the Death with Dignity Act that emanated therefrom. The fact that the U.S. Supreme Court upheld the law on procedural grounds (relating to whether or not it was unlawful for physicians to write prescriptions for a lethal dose, under federal law, or permissible as a matter of state law), rather than on the substantive question of whether assisted suicide should be allowed propelled proponents of assisted suicide, and repelled opponents (including dissent in the U.S. Supreme Court). The 2006 U.S. Supreme Court, upholding the Oregon law and its protocols, laid the groundwork for further legislation in Washington State, and for further litigation.

State supreme courts are split in their treatment of physician-assisted suicide (but not medical euthanasia, which is banned and criminal throughout the country, as is mercy killing). The Michigan Supreme Court, in response to Kevorkian's activities and in response to the very first civil challenge regarding assisted suicide, determined that assisted suicide was illegal. While the little-discussed (outside of Michigan) case of Teresa Hobbins (et al.) also was considered, and ruled against by the Michigan Supreme Court, there was no doubt by any of the major (or minor) players in 1990s Michigan that, in no small measure, Kevorkian's increasing, escalating, media-seeking, gendered activities had a role in the court's decision-making process.

On the other hand, the Montana Supreme Court, perhaps with the hindsight visual benefits of the Oregon and Washington State experiences, as well as the U.S. Supreme Court positions in 1997 and 2006, ruled that as a matter of state constitutional law, physician-assisted suicide should be permitted. The question is whether this reflects a sea change in the legal culture, or a view of the medical culture. In either event, this collection of U.S. Supreme Court and state supreme court cases shows a distinct curve toward permission, rather than prohibition, of physician-assisted suicide. This last statement is not to be considered an endorsement (or the opposite) of physician-assisted suicide, but merely an observation pertaining to what appears to be a trend.

THE WAY FORWARD: A VIEW AND PERSPECTIVE OF THE PAST

A view of the social and legal constructions of medical euthanasia and physician-assisted suicide offers one definitive conclusion at the time of this writing. The legal theorists of the 1950s—Kamisar, Williams, and St. John-Stevas—were all right to some degree in their approaches, and one can look to their work for a perspective of the applications of social and legal approaches to medical euthanasia and physician-assisted suicide in the 2000s. New Orleans' Dr. Pou benefitted from Kamisar's law in action as being as malleable as the law on the books was not. Montana's Supreme Court ruling in as to how Baxter, patients, and doctors should be treated as regards assisted suicide (admittedly not a concept that was a gleam in anyone's socio-legal eye in the 1950s) was perhaps best viewed through Williams's eyes. St. John-Stevas (and Kamisar and Williams) might well have endorsed the way that Georgia's Carol Carr was treated after she shot her terminally ill sons.

While none of the original trio would likely have contemplated an Oregon or a Washington, neither would they have considered the possibility of a Kevorkian. These appear to have shown the best and the worst of the possibilities for the assisted-dying debate in the United States, from a legislative and implementation standpoint (and, as to Kevorkian, whose appeal was denied in the new millennium, continued litigation).

There will be developments, doubtless on both sides—nay, all sides—of these issues as medical technologies and abilities further emerge, as social and legal data continue to be collected from states where assisted suicide is permitted, and with regard to cases of medical euthanasia that emerge (whether from seeming necessity, duress, or otherwise). Ultimately, the systems of values were, and may never be, completely internally consistent (let alone externally so), as regards the hastening of the death of a willing, indeed requesting, patient, and these may not resoluble. As stated in the Introduction to this book, and revisited after considering the various chapters and decades, Michael Ignatieff, writing the biographical *A Life: Isaiah Berlin*, put it well:

> Systems of values were never internally consistent. The conflict of values—liberty versus mercy; tolerance versus order; liberty versus social justice; resistance versus prudence—was intrinsic to human life.[1]

In matters of assisted death, there are discrepant goods that are not complementary, and the attempts to litigate, legislate, and relitigate made them neither compatible nor resolvable. Perhaps the best answer to how the debate should or will be resolved in the future, is that it has had difficult and inconsistent resolutions in the past. Perhaps that is the greatest value of a historical perspective—to be able to find the order in the chaos of organic growth in opposing directions.

NOTE

1. Igantieff, Michael, *A Life: Isaiah Berlin* (New York: Metropolitan Books, 1998), 285.

Appendix

Table of Cases, Legislation, Proposed Legislation, and Initiatives

It is a convention in legal briefs to construct a Table of Cases, Legislation and (if appropriate) Legislative Proposals and Initiatives. There is also a convention of having all cites to all sources in one list.

For this treatment of the euthanasia and assisted-suicide debate, there is a reference list of sources cited, and there is also this list. The latter provided some challenge, best explained at the outset. Some cases were retrieved from news sources (especially trials to acquittal, which have no official record or which have records which are sealed, Lord Devlin's memoir of the Bodkin Adams case aside, and the sole mysteriously extant copy of the original trial transcript, of uncertain but authentic origin, at the Institute for Advanced Legal Studies in London), some (specifically the final 1999 Kevorkian trial to conviction) had a trial transcript, some appellate matters and matters for declaratory judgment were reported. Reported cases have a system of reporting, but unreported cases and transcripts, but more complicated is the collection of unreported cases, for which case numbers are being used. For those obtained out of news (for example, Bodkin Adams and Hermann Sander in the 1950s), State, Case Name, News Reports as Cited in Text, along with date. This is an approximation of the system of case reporting for officially reported cases.

With legislation (and proposals/initiatives), for ease of the reader or reference seeker, state names are also listed first.

Baxter v. State, 354 Mont. 234 (Mont. 2009).

California Proposition 161, the Aid-in-Dying Act (1992).

Compassion in Dying v. State, 850 F. Supp. 1454 (W.D.Wa, 1994).

Compassion in Dying v. State of Washington, 49 F.3d 586, 589 (9th Cir. 1995).

Connecticut Act to Legalize Euthanasia (1959).

Cruzan v. Director, Missouri Department of Health, 497 U.S. 261 (1990).

Cruzan v. Harmon, 760 S.W.2d 408, 411 (1988).

(England and Wales) The Suicide Act 1961.

(England) 1936 Bill The Voluntary Euthanasia (Legislation) Bill.

Georgia "Offering to Assist in the Commission of a Suicide," Code 1981, § 16-5-5, enacted by Ga. L. 1994, p. 1370, § 1; Ga. L. 2007, p. 133, § 5/HB 24.

Gonzales v. Oregon, 546 U.S. 243 (2006).

Hobbins v. Attorney General, No. 93-306-178 CZ (Cir. Ct. Wayne Cty., May 20, 1993) (Stephens, J.) (unreported decision).

In re Quinlan, 70 N.J. 10 (Sup. Ct. N.J., 1976).

In re Quinlan, 173 N.J. 227, 237 (1975).

Iowa Proposal for Euthanasia (1906).

Lee v. Oregon, 107 F.3d 1382 (3d Cir. 1997).

Michigan Code Archive Directory, Crimes and Offenses, Assistance to Suicide, Section 752.1027 (1993).

Michigan House (HB 4038)/Michigan Senate (SB 32) Crimes; Homicide' Aided Suicides, Prohibit (1991).

Michigan Proposal B (1998). The Terminally Ill Patient's Right to End Unbearable Pain or Suffering (failed).

Michigan Public Act 270 (1992), Michigan Public Act 3 (1993) (immediate effect provision). An Act to create the Michigan commission on death and dying; to prescribe its membership, powers, and duties; to provide for the development of legislative recommendations concerning certain issues related to death and dying; to prohibit certain acts pertaining to the assistance of suicide; to prescribe penalties; and to repeal certain parts of this act on a specific date.

Nebraska Voluntary Euthanasia Act (Proposed) (1937).

New York Penal Law Section 125.15 (3).

New York Penal Law Section 120.30.

Ohio General Assembly Proposal for Euthanasia (1906).

Oregon Death with Dignity Act, ORS 127.800 *et. seq.* (1995).

Oregon Measure 16, The Death with Dignity Act (1994).

People v. Campbell, 124 Mich.App. 333 (Mich. Ct. App. 1983).

People v. Carol Carr (Georgia 2002) (as reported in press).

People v. Hermann Sander (New Hampshire, 1950) (as reported in press).

People v. Kevorkian, 248 Mich.App. 373 (2001).

Original (and Complete) Copy of Trial Transcript and Sentencing Proceedings, *People v. Kevorkian*, No 98-163675-PC (March-April 1999).

March 22, 1999 (Volume 2 of 4)
March 25, 1999 (Volume 3 of 4)
April 13, 1999 (Sentencing Proceedings)

People v. Kevorkian, 447 Mich. 436 (1994).

People v. Kevorkian, 205 Mich.App. 180 (Court of Appeals of Michigan, May 10, 1994), reversing, Case Number 92-115190-FC.

People v. Kevorkian, Case Number 93-129832-FH and Case Number 94-130248-FH (Cir. Ct. Oakland Cty., January 27, 1994) (Cooper, J. writing) (unreported decision).

People v. Kevorkian, Case No-CR92-115190-FC and 92 –DA-5303-AR, Opinion and Order (Cir. Ct. Oakland County, July 21, 1992) (Breck, J, writing), p. 12 (unreported decision).

People v. Anna Pou (Louisiana, 2007) (as reported in press).

People v. Roberts, 211 Mich. 187 (1920).

People v. Timothy Quill (New York, 1991) (as reported in press).

Quill v. Koppell, 870 F. Supp. 78 (S.D.N.Y. 1994).

R. (on the application of Purdy) (Appellant v. Director of Public Prosecutions (Respondent), [2009] UKHL 45.

Regina v. Harold Shipman (England, 2002) (as reported in press).

Regina v. John Bodkin Adams. Transcript of the shorthand notes of Geo Walpole and Co., (Official Shorthand writers to the Centra Criminal Court), "Seventeenth Day," "Summing up," Tuesday April 9, 1957.

Regina v. John Bodkin Adams (England, 1957) (as reported by Lord Devlin).

Regina v. Nigel Cox (England, 1992) (as reported in press).

Roe v. Wade, 410 U.S. 113 (1973).

Vacco v. Quill, 521 U.S. 793 (1997).

Washington Aid-in-Dying, Initiative 119 (1991).

Washington Death with Dignity Act, RCW 70.245 *et seq*. (2009).

Washington Rev. Code Section 9A.36.060 (promoting a suicide attempt).

Washington v. Glucksberg, 521 U.S. 702 (1997).

Glossary

Active euthanasia. Generally refers to a positive act by someone other than the patient, which is intended to bring about the death of the patient, an example of which is the injection of a lethal injection.

Advance directive. A document by which a person makes known their wishes that they want no further treatment, in the event of a catastrophic accident or life-threatening condition that precludes them from making their own rational decisions (or preempts their legal capacity). More expansive than a living will in that an advance directive may also provide for directions to name a health care proxy (or agent).

Artificial life support. Treatment or procedures used to maintain life after one or more vital organs fail (for example, due to trauma from an accident or a heart attack or chronic or degenerative illness.

Assisted suicide. When one person helps another person commit suicide (end their life), usually by providing the means by which a person ends their life.

Autonomy. Respect for the choices of a competent patient in matters regarding health care and choices which might shorten or lengthen life.

Brain dead. Where there is irreversible cessation of brain activity, usually following the loss of oxygen to the brain.

Degenerative illness. A reference to an ongoing, irreversible illness, such as ALS (Lou Gehrig's disease), Huntington's disease or multiple sclerosis, which will inevitably result in death.

Double effect. Where the patient's treatment has as its primary purpose pain relief and/or palliation and/or relief of discomfort, and has as a secondary effect the shortening of life.

Euthanasia. The acceleration, causing or hastening of death, particularly where the patient is terminally ill or incurably ill with a degenerative illness, with the final life-shortening act performed by somebody other than the patient. The two most basic forms are active and passive.

Hospice. Philosophy that the dying need to be provided with end-of-life care (for example, palliative care) to address physical, emotional or social needs of the patient. Also treatment consistent with this philosophy.

Involuntary euthanasia. Where euthanasia is administered to someone who neither consents to nor requests death (for example, people who were administered euthanasia in the course of the Nazi atrocities).

Living will. A limited type of advance directive by which a person makes decisions about life-sustaining treatment and/or procedures, in the event that their death from a terminal condition is imminent (regardless of life-sustaining treatment) or if she or he is in a persistent vegetative state (permanent unconsciousness).

Medical aid in dying. A term that includes both the practice of euthanasia (as performed by members of the health care team) and assisted suicide (usually as performed by physicians).

Mercy killing. Nonmedical, compassionate hastening of the death of someone with a terminal illness or chronic degenerative condition, usually for a compassionate or a humanitarian reason (for example, where a man or woman shoots or suffocates or administers poison to a spouse, friend, parent, or child).

Natural death. Generally refers to dying without medical or technological intervention, or after artificial life support is withdrawn (for example, by infection or pneumonia), so as to let nature take its own course. Phrasing relates to death by natural causes (for example, by a heart attack).

Non-voluntary euthanasia. Where euthanasia is performed without the patient's consent (for example, where a patient has no capacity due to a degenerative condition, such as where the patient is in a persistent vegetative state).

Palliative care. Where treatment focuses upon reducing (or mitigating) the severity of symptoms of a disease, rather than to try to cure or slow the progression of a disease, also provision of end of live comfort care.

Passive euthanasia. Denotes an act by which a patient's treatment, nutrition, and/or hydration are discontinued; the cessation of life-sustaining treatment.

Persistent vegetative state. Generally a person who is in an ongoing coma who can be kept alive indefinitely by the use of medical technology, and where removing the medical technology would result in inevitable natural death.

Physician-assisted suicide. Differs from euthanasia in that it is the patient, rather than the doctor, who commits the life-shortening act, although the physician provides the means (such as where a doctor prescribes lethal doses of painkillers or barbiturates, which the patient then takes orally).

Right to die. Regards the rejection of life-sustaining treatment, with an emphasis on life-prolonging technology (for example, artificial nutrition and hydration or respirator). Sometimes referred to as the right to refuse medical treatment.

Unnatural death. Usually refers to death by accident, suicide, or homicide.

Voluntary euthanasia. Where the patient expressly asks to have euthanasia administered (for example, a patient at the end stage of an illness who is of age and capacity to consent, who expressly requests death).

Bibliography

Appel, Jacob M. "A Duty to Kill? A Duty to Die? Rethinking the Euthanasia Controversy of 1906." *Bulletin of the History of Medicine* 78, no. 3 (2004): 610-34. http://search.proquest.com/docview/236572724?accountid=14055.

Atlanta Journal-Constitution.

Bachman, Jerald G., Kristen H. Alcser, David J. Doukas, Richard L. Lichtenstein, and Amy D. Corning. "Attitudes of Michigan Physicians and the Public toward Legalizing Physician Assisted Suicide and Voluntary Euthanasia." *New England Journal of Medicine* 334 (February 1, 1996): 303–9.

Barrington, Mary R. "Euthanasia: An English Perspective." In *To Die or Not to Die: Cross-Disciplinary, Cultural and Legal Perspectives on the Right to Choose Death.* New York: Praeger, 1990.

Beedle, Cameron (Kevorkian trial juror, April–May 1996). Interview by author. Michigan, May 19, 1996.

Breck, David (Kevorkian trial judge, April–May 1996). Interview by author. Pontiac, Michigan, May 15, 1996.

British Medical Association. *Euthanasia.* London, 1988.

Burleigh, Michael. *Death and Deliverance: "Euthanasia" in Germany 1900–1945.* Cambridge: Cambridge University Press, 1994.

Caplan, Arthur L. *When Medicine Went Mad: Bioethics and the Holocaust.* Totowa, NJ: Humana Press, 1992.

Chin, Arthur E., Katrina Hedberg, Grant K. Higginson, and David W. Fleming. "Legalized Physician-Assisted Suicide in Oregon—the First Year's Experience." *New England Journal of Medicine* 340 (February 18, 1999): 577–83.

Ciamaritaro, Nick (Michigan state legislator). Interview by author. Lansing, Michigan, March 2, 1994.

Cohen, Stanley. *Folk Devils and Moral Panics: The Creation of the Mods and the Rockers*. 2nd ed. Oxford: Blackwell Publications, 1987.

Cooper, Jessica (Kevorkian trial judge, March 1996 and 1999). Interview by author. Michigan, May 17, 1996.

Death with Dignity Act 2011 Annual Report. Oregon Public Health Division. http://public.health.oregon.gov/ProviderPartnerResources/EvaluationResearch/. Accessed March 18, 2012.

Devlin, John P. *Easing the Passing: The Trial of Dr. John Bodkin Adams*. London: Faber and Faber, 1985.

Dillingham, Fred (Michigan state senator). Interview by author. Michigan, August 23, 1993.

Durkheim, Emile. *Suicide: A Study in Sociology*. New York: Free Press, 1966.

Emanuel, Ezekiel J. "The History of the Euthanasia Debates in the United States and Britain." *Annals of Internal Medicine* 121, no. 10 (November 15, 1994): 793–802.

Fletcher, George P. *Rethinking Criminal Law*. Oxford: Oxford University Press, 2001.

A Group of People (also known as the Final Report of the Michigan Commission on Death and Dying). 1994.

Guardian (London).

Hawkins, J. M., ed. *The Oxford Paperback Dictionary*. 3rd ed. Oxford: Oxford University Press, 1990.

Hendin, Herbert, and Kathleen Foley. "Physician-Assisted Suicide in Oregon: A Medical Perspective." *Michigan Law Review* 106 (2008): 1613–40.

Herring, Mark Y. *Historical Guides to Controversial Issues in America: The Pro-Life/Choice Debate*. Westport, CT: Greenwood Press, 2003.

Hilliard, Bryan. "The Moral and Legal Status of Physician-Assisted Death: Quality of Life and the Patient Physician Relationship." *Issues in Integrative Studies* 18 (2000): 45–63.

Hilliard, Daniel, and John Dombrink. *Dying Right: The Death with Dignity Movement*. New York: Routledge, 2002.

Hoefler, James M., and Brian E. Kamoie. *Deathright: Culture, Medicine, Politics and the Right to Die*. Boulder, CO: Westview Press, 1994.

House of Lords. *Oral Evidence, House of Lords Select Committee on Medical Ethics, 1993–1994* (1994).

House of Lords. *Report, House of Lords Select Committee on Medical Ethics, 1993–1994* (1994).

House of Lords, Select Committee on Medical Ethics. "Medical Ethics: Select Committee Report." May 9, 1994. http://hansard.millbanksystems.com/lords/1994/may/09/medical-ethics-select-committee-report. Accessed February 2, 2011.

Howarth, Glennys. *Death and Dying: A Sociological Introduction*. Cambridge: Polity Press, 2007.

Howarth, Glennys. "Last Rites: The Work of the Modern Funeral Director." PhD diss., London School of Economics and Political Science, 1991.

How to Die in Oregon. Directed by Peter D. Richardson. Performed by Documentary. HBO Documentaries, 2011.

Humphry, Derek. *Final Exit: The Practicalities of Self-Deliverance and Assisted Suicide for the Dying*. Eugene, OR: Hemlock Society, 1991.

Humphry, Derek, and Ann Wickett. *The Right to Die: Understanding Euthanasia*. New York: Harper & Row, 1986.

Ignatieff, Michael. *Isaiah Berlin: A Life*. New York: Metropolitan Books, 1998.

Jennett, Bryan, and Fred Plum. "Persistent Vegetative State after Brain Damage: A Syndrome in Search of a Name." *The Lancet* 7553: 734–37.

Kamisar, Yale. "Euthanasia Legislation: Some Non-Religious Objections." In *Voluntary Euthanasia: Experts Debate the Right to Die*, 110–55. London: Peter Owen, 1986.

Kamisar, Yale. "Some Non-Religious Views Against Proposed 'Mercy Killing' Legislation." *Minnesota Law Review* 42, no. 6 (May 1958): 969–1042.

Kearl, Michael C. *Endings: A Sociology of Death and Dying*. Oxford: Oxford University Press, 1989.

Kemp, N. D. A. *"Merciful Release": The History of the British Euthanasia Movement*. Manchester: Manchester University Press, 2002.

Keown, John. *Euthanasia, Ethics and Public Policy: An Argument against Legalisation*. Cambridge: Cambridge University Press, 1997.

Kuepper, Stephen L. "Euthanasia in America, 1890–1960: The Controversy, the Movement and the Law." PhD diss., Rutgers University, the State University of New Jersey at New Brunswick, 1981.

Lamb, David. *Down the Slippery Slope: Arguments in Applied Ethics*. London: Croon Helm, 1988.

Lelyveld, Joseph. "1946 Secret Is Out: Doctor Sped George V's Death." *New York Times*, November 27, 1986.

McGough, Peter M. "Washington State Initiative 119: The First Public Vote on Legalizing Physician Assisted Death." *Cambridge Quarterly of Healthcare Ethics* 2 (1992): 63–67.

Miel, Charles (Kevorkian trial judge, 1997). Personal interview, Ionia, MI. June 12, 1997.

Miner, Horace. "Body Ritual among the Nacirema." *American Anthropologist* 58, no. 3 (1956): 503–7.

New York State Task Force on Life and the Law. *When Death Is Sought: Assisted Suicide and Euthanasia in Context*. New York: Self-published, 1994.

New York Times.

Nichol, Neal, and Harry Wylie. *Between Dying and the Dead: Dr. Jack Kevorkian's Life and the Battle to Legalize Euthanasia*. Madison: University of Wisconsin Press, 2006.

Ogden, Russell. *Euthanasia and Assisted Suicide in Persons with Acquired Immunodeficiency Syndrome (AIDS) or Human Immunodeficiency Virus (HIV)*. New Westminster, British Columbia: Peroglyphics Publishing, 1994.

Oregonian.

Pakes, Francis. *Comparative Criminal Justice.* Devon: Willan Publishing, 2004.

Pappas, Demetra M. "The Politics of Euthanasia and Assisted Suicide: A Case Study of the Emerging Criminal Law and the Criminal Trials of Jack 'Dr. Death' Kevorkian." PhD diss., London School of Economics and Political Science, 2009.

Pappas, Demetra M. "Recent Historical Perspectives Regarding Medical Euthanasia and Physician Assisted Suicide." *British Medical Bulletin* 52, no. 2 (1996): 396–93.

Post, Stephen G. "A Postmorten on Initiative 119." *Health Progress,* January–February 1992, 70.

Pross, Christian. "Nazi Doctors, German Medicine and Historical Truth." In *The Nazi Doctors and the Nuremberg Code: Human Rights in Human Experimentation.* New York: Oxford University Press, 1992.

Quill, Timothy. *Death and Dignity: Making Choices and Taking Charge.* New York: W. W. Norton & Company, 1993.

Quill, Timothy. Foreword. In *Hospice or Hemlock: Searching for Heroic Compassion,* ix–xii. Westport, CT: Praeger, 2002.

Quill, Timothy E. "Sounding Board, Death and Dignity: A Case of Individualized Decision Making." *New England Journal of Medicine* 325, no. 10 (March 7, 1991): 691–94.

Rosco, Lori A., L. J. Dragovic, and Donna Cohen. "Dr. Jack Kevorkian and Cases of Euthanasia in Oakland County, Michigan, 1990–1998." *New England Journal of Medicine* 343 (2000): 1735–36.

Sanford, John (member, Michigan Commission on Death and Dying). Interview by author, Michigan, March 1994.

The Sea Inside. Directed by Alejandro Amenabar. Performed by Javier Bardem. Fine Line Features, 2004. Film.

St. John-Stevas, Norman. *Life, Death and the Law: Law and Christian Morals in England and the United States.* Bloomington: Indiana University Press, 1961.

Strauss, Anselm L., and Barney G. Glaser. *Anguish: A Case History of a Dying Trajectory.* San Francisco: University of California Medical Center, 1970.

Sullivan, Amy D., Katrina Hedberg, and David W. Fleming. "Legalized Physician-Assisted Suicide in Oregon—the Second Year." *New England Journal of Medicine* 342 (February 24, 2000): 598–604

Swarte, Nikkie B., Marije L. Van Der Lee, Johanna G. Van Der Bom, Jan Van Den Bout, and A. Peter M. Heintz. "Effects of Euthanasia on the Bereaved Family and Friends: A Cross-Sectional Study." *British Medical Journal,* no. 7408 (July 26, 2003): 189.

Taylor, Steve and Field, D., eds. *The Sociology of Health and Health Care: An Introduction for Nurses.* Oxford: Blackwell, 1993.

Thompson, William E. "Handling the Stigma of Handling the Dead: Morticians and Funeral Directors." *Deviant Behavior: An Interdisciplinary Journal* 12 (1991): 403–29.

Voet, Ray (chief prosecuting attorney; Ionia County 1997 Kevorkian trial lawyer). Interview by author, Ionia, MI. June 13, 1997.

Volker, Deborah L. "The Oregon Experience with Assisted Suicide." *Journal of Nursing Law* 11, no. 3 (2005): 152–62.

Wallace, Mike, writer and moderator. "Death by Doctor." On *60 Minutes*. CBS. November 22, 1998.

Watson, Francis. "The Death of George V." *History Today* 36, no. 12 (December 1986): 21–31.

Williams, Glanville. *The Sanctity of Life and the Criminal Law.* New York: Alfred A. Knopf, 1957.

Wolfson, Jay. "Foreword." In *The Case of Terri Schiavo: Ethics at the End of Life*, edited by Arthur L. Caplan, James J. McCartney, and Dominic A. Sisti. Amherst, NY: Prometheus Books, 2006.

You Don't Know Jack. Directed by Barry Levinson. Performed by Al Pacino. HBO Films, 2010.

Zalman, Marvin, John Strate, Dennis Hunter, and James Sellars. "The Michigan Assisted Suicide Three Ring Circus." *Ohio Northern University Law Review* 23, no. 3 (1997): 863–968.

Index

About the Author

DEMETRA M. PAPPAS, JD, MSc, PhD, holds a law degree from Fordham University School of Law. After engaging in public interest work as a trial and appellate criminal defense attorney for the Legal Aid Society, and completing a judicial clerkship to an appellate judge in New York, she returned to graduate school where she achieved an MSc in Criminal Justice Policy from the London School of Economics and Political Science, and a PhD from the LSE. In 2009, she defended her doctoral work, "The Politics of Euthanasia and Assisted Suicide: A Comparative Case Study of Emerging Criminal Law and the Criminal Trials of Jack 'Dr. Death' Kevorkian," at the LSE (Departments of Law and Sociology). In addition to her work on end-of-life issues, she writes for both academic and general readerships about a number of other subjects, including emerging legislation, criminal law, criminology, sociological representations of celebrity and the media, sociology of medicine, dramaturgical representations in the sociological and legal contexts, culinary culture/food studies, visual sociology and pedagogical methodology, among others. She also writes about anti-stalking mechanisms; for one of her articles on this topic, she earned the 1997 LSE William Robson Memorial Prize for her ethnographic study in Minnesota. Most recently, she was awarded the SGA (Student Government Association) Faculty Member of the Year Award 2011–2012 at St. Francis College, for her teaching in the Department of Sociology and Criminal Justice. She may be reached by e-mail at DemetraPappas@yahoo.com and followed on Twitter @DemetraPappas.